FLUID POWER SYSTEMS
Theory, worked examples and problems

Pg 31 Nos 1 & 2 & 3
73 No 1

FLUID POWER SYSTEMS
Theory, worked examples and problems

A. B. GOODWIN
Principal Lecturer in
Mechanical Engineering
Leicester Polytechnic

First published 1976 by

THE MACMILLAN PRESS LTD

London and Basingstoke

Associated companies in New York Dublin

Melbourne Johannesburg and Madras

SBN 333 19368 7

Printed in Great Britain by
Thomson Litho Limited., East Kilbride, Scotland.

D
620.106
Goo

CONTENTS

PREFACE

Many books have been written on the subject of fluid power
since the well-known works of Blackburn, Reethof and
Shearer. Many of these were inspired by research, or pro-
jects, undertaken by their authors. As a teacher I have
written this book with the needs of students in mind. My
aim was to provide a text giving a learning-by-example
treatment of the various systems used in fluid power.
Basic theory is established in each chapter and selected
worked examples cover many particular points in detail.
Many additional problems are included for the reader to
solve and these may be selected for H.N.C., H.N.D.,
C.E.I.2 and B.Sc. levels. Post-graduate students,
specialising in this area of work, may also find the book
of considerable help in their studies.

I have started with basic hydraulic components as sub-
systems and defined them in terms of their steady-state
characteristics. These components are then linked to form
various systems and analysed for steady-state and transient response under various loading conditions.

The chapter dealing with feedback systems is confined
to the derivation of system equations and transfer functions, their physical interpretation and graphical representation. Response and stability criteria have been
omitted since the general theory for this study is well
expounded in many other books. With a knowledge of basic
control theory and the techniques of Routh, Nyquist, Bode,
Nichols and methods of improving system performance, an
even deeper study of the behaviour of fluid power systems
can be made.

My thanks are expressed to my colleagues at Leicester
and to my students whose help and tolerance made this book
possible.

Bernard Goodwin

1   BASIC FLUID POWER COMPONENTS

DEFINITIONS USING STEADY-STATE CHARACTERISTICS

Pump

This circulates the fluid around the system and provides
the pressure necessary to overcome the load at its outlet
port. All pumps used in hydrostatic systems are positive-
displacement, i.e. a definite volume of pumping chamber
is swept out for every revolution of the drive shaft.
This volume is the capacity of the pump, which may be
fixed capacity, i.e. the pump is constructed so that the
capacity is constant; or variable capacity, i.e. the
pump construction is such that the capacity may be varied
by suitable adjustment of the stroke mechanism. Note:
although the capacity may be varied, the unit is still
positive-displacement.

Examples

fixed capacity: gear, piston, vane pumps
variable capacity: commonly axial piston and vane pumps.

    For steady-state conditions, with ideal pumps, the
delivery (volume per unit time) Q_p is given by

$$Q_p = C_p n_p (\text{or } C_p \omega_p) \tag{1.1}$$

where C_p = pump capacity per rev. or rad. of shaft

      n_p = pump shaft speed revolutions per unit time

      ω_p = pump shaft speed radians per unit time (= $2\pi n_p$)

e.g. $Q_p(\text{ml/s}) = C_p(\text{ml/rad}) \times \omega_p(\text{rad/s})$

    No pump is perfect, hence there will always be a leak-
age of oil, under pressure, through the clearance spaces
in the unit and there will be friction between moving
surfaces. The leakage will cause the delivered volume
per unit time to be less than that obtained by eqn 1.1.
Because of the dimensions of clearance spaces it is
reasonable to assume laminar flow within them, so the
leakage rate may be expressed in the form

$$Q_1 = \frac{k \, \Delta p}{\mu} \tag{1.2}$$

where
      Q_1 = leakage rate

Δp = pressure difference across leakage path (frequently assumed to be pump delivery pressure)
μ = fluid dynamic viscosity
k = a constant for the particular pump

At constant temperature $k/\mu = \lambda_p$, the pump leakage coefficient.

The mechanical losses within the pump will result in a, power input to the total pumped oil being less than the shaft input to the pump.

The total oil flow rate through the pump will be $C_p\omega_p$ and all of this will reach the pressure Δp_p above inlet pressure, hence the input to the oil will be $C_p\omega_p\Delta p_p$ even though Q_1 is leakage.

The shaft power input will be $T_p\omega_p$ where

T_p = pump shaft torque
ω_p = pump shaft speed (rad/s)

The relationship between these powers is the mechanical efficiency of the pump (sometimes called torque efficiency) $_p\eta_m$.

$$_p\eta_m = \frac{C_p\,\Delta p_p}{T_p} \tag{1.3}$$

(Remember that C_p is the swept volume per radian of shaft rotation.)

The overall efficiency of the pump is the ratio of power in delivered fluid to shaft power

$$_p\eta_o = \frac{(Q_p - Q_1)\Delta p_p}{T_p\omega_p} \tag{1.4}$$

The volumetric efficiency may be expressed as the ratio of delivery rate to swept rate

$$_p\eta_v = \frac{Q_p - Q_1}{Q_p} \tag{1.5}$$

Note that eqn 1.3 × 1.5 gives

$$_p\eta_m \times {}_p\eta_v = \frac{C_p\,\Delta p_p}{T_p} \times \frac{Q_p - Q_1}{Q_p}$$

but since $Q_p = C_p\omega_p$

2

$$p^{\eta}m \times p^{\eta}v = \frac{(Q_p - Q_1)\Delta p_p}{T_p \omega_p} \qquad (1.6)$$

Note: The efficiencies as defined here are instantaneous efficiencies, i.e. they are true for one set of conditions only.

Fig. 1.1 indicates the steady-state conditions for a positive-displacement pump.

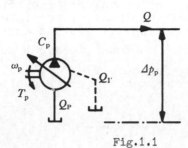

Fig.1.1

Relief Valve

The function of the relief valve is to limit the maximum pressure that can exist in a system. Under ideal conditions the relief valve should provide an alternative path to tank for the system oil while keeping the maximum pressure constant. This implies that the pressure should not vary with the quantity of oil passing through the relief valve to tank. Hence the ideal relief valve is defined by pressure alone. Referring to fig. 1.2

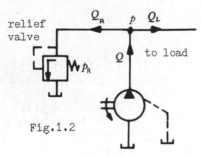

Fig.1.2

if $\quad p < p_R$ then $Q_L = Q$

and if

$\quad\quad p = p_R$ then $Q_R = Q - Q_L$

also $p \ngtr p_R$

3

However, a relief valve is merely an orifice through which the oil passes and is governed by the normal orifice flow formula

$$Q_R = ka_o\sqrt{\Delta p}$$

where a_o is the orifice area and Δp the pressure drop across it. For a relief valve $\Delta p = p_R$ since the discharge is to tank.

For a direct-acting relief valve the orifice is opened when the pressure p of the system exceeds the pressure value of the spring p_R. For an increased flow Q_R the orifice area must increase, hence p must increase to compress the spring further (fig. 1.3). If it is assumed that the orifice area, a_o, is proportional to

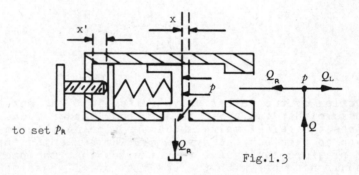

to set p_R

Fig.1.3

spool displacement x and that x is proportional to $(p - p_R)$, i.e. $x = 0$ at $p = p_R$, then, in the steady state

$$Q_R = K(p - p_R)\sqrt{p} \qquad (1.7)$$

High-pressure relief valves (i.e. p_R high) will require high-stiffness springs in order that the initial precompression is not excessive (x'). Since K in eqn 1.7 is inversely proportional to the spring stiffness then a high stiffness will result in a high value of $(p - p_R)$ when Q_R is large. Note that $(p - p_R)$ is sometimes called the pressure over-ride.

Eqn 1.7 is true up to the point at which the relief valve is fully open. Let this be at pressure p_1. When this condition is reached the port area cannot increase and the valve becomes a fixed area orifice with the flow eqn

$$Q_R = K(p_1 - p_R)\sqrt{p} \qquad (1.8)$$

4

Fig. 1.4 is a plot of eqns 1.7 and 1.8 showing the useful part of the characteristic up to pressure p_1. The angle θ is $\tan^{-1}(p - p_R)/Q_R$, i.e. $\tan \theta$ is approximately proportional to K, hence the greater the spring stiffness the larger is θ and the pressure over-ride.

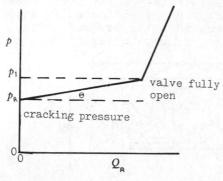

Fig.1.4

Non-return Valve

A non-return valve is similar in construction to a direct-acting relief valve but it is placed in series with the pump flow, whereas a relief valve is in parallel.

In fig. 1.5, Δp is the pressure difference across the valve, Δp_N is the cracking pressure and Δp_1 is the value

Fig.1.5

of Δp when the non-return valve is fully open. By suitable selection of spring stiffness and Δp_N, the system pressure upstream of the non-return valve can be kept above a minimum value Δp.

For the non-return valve

$$Q_N = K(\Delta p - \Delta p_N)\sqrt{\Delta p} \qquad (1.9)$$

and at maximum valve opening

$$Q_N = K(\Delta p_1 - \Delta p_N)\sqrt{\Delta p} \qquad (1.10)$$

The similarity between eqns 1.7, 1.8, 1.9 and 1.10 can

be seen and fig. 1.4 represents the steady state of a
non-return valve with Δp, etc. substituted for p, etc.

Pilot-operated Relief Valve

In order to reduce the pressure over-ride of a relief
valve a pilot stage is frequently introduced. The prin-
ciple of this operation is indicated in fig. 1.6. The
pilot valve A opens when the pressure acting on it
exceeds the setting of the spring B; let this be p_R.

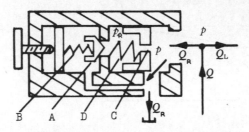

Fig.1.6

The resulting small flow of oil through the orifice C
causes main pressure p to exceed p_R. It is so arranged
that this resulting pressure difference will overcome
the main spool spring D and allow oil to dump to tank.
The spring B sets the nominal relief pressure (high-
stiffness spring) and spring D merely keeps the main
spool port closed (low-stiffness spring).

    The flow equation for the main spool orifice is the
same as eqn 1.7 with the exception that p_R is not provided
by pre-compression of the spring (x' in fig. 1.3) but
is provided by the oil. This means that the spring D
is of small stiffness and so the angle θ in fig. 1.4
is small, giving very little pressure over-ride.
The point p_1 (valve fully open) is also at a higher
value of Q_R since more spring compression is available
due to the absence of pre-compression.

Pipes

These form the connecting links between hydraulic components and, in the steady state, have a fixed physical
size. If the fluid is taken in association with the
pipe then two types of flow are usually considered.

    (a) Laminar flow - where the pressure loss in a given
length is proportional to the flow rate, i.e. $\Delta p \propto Q$.
    (b) Turbulent flow - where the pressure loss in a
given length is proportional to the square of the flow
rate, i.e. $\Delta p \propto Q^2$.

6

For straight circular pipes

Laminar flow

$$\Delta p = \frac{32\mu lv}{d^2} \tag{1.11}$$

or $$\quad \Delta p = \frac{128\mu lQ}{\pi d^4} \tag{1.11}$$

where Δp = pressure loss
    v = mean velocity of flow
    Q = flow rate = $v \times \pi d^2/4$
    d = pipe diameter
    l = pipe length
    μ = fluid dynamic viscosity

This relationship may be considered correct up to a value of Reynolds number of 1200. Between 1200 and 2500 is a transition region where the flow changes to turbulent. Above 2500 the flow is fully turbulent. Reynolds number Re = $vd\rho/\mu$ where ρ = fluid density.

Turbulent flow

$$\Delta p = \frac{4fl\rho v^2}{2d} \tag{1.12}$$

where f is a pipe friction factor determined experimentally. It must be noted that 4f is used in many hydraulic references to replace f above, but it is still referred to as f, hence

$$\Delta p = \frac{fl\rho v^2}{2d} \tag{1.13}$$

This point must be watched very carefully.

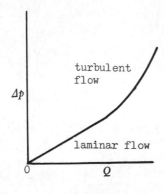

Fig.1.7a

7

Now f in eqn 1.12 can be taken as $f = 0.08(Re)^{-0.25}$, for Re between 2500 and 10^5 and similarly f in eqn 1.13 is given by $f = 0.32(Re)^{-0.25}$. Eqns 1.12 and 1.13 may also be used for laminar flow conditions if, in eqn 1.12 $f = 16/Re$ and in eqn 1.13 $f = 64/Re$.

Fig. 1.7a indicates the steady-state characteristic for a pipe and fig. 1.7b shows some typical values.

Notes on units

(1) Dynamic viscosity has dimensions (M/LT), hence the SI unit is kg m^{-1} s^{-1}. The existing unit of 1 centipoise (1cP) = 1 g cm^{-1} s^{-1} $\times 10^{-2}$ = 1 kg m^{-1} s^{-1} $\times 10^{-3}$, i.e. 1000 cP = 1 SI unit (unnamed).

(2) Kinematic viscosity ($\nu = \mu/\rho$) has dimensions (L^2T^{-1}) hence the SI unit is m^2 s^{-1}. 1 centistoke (1cSt) = 1 cm^2 s^{-1} = 10^{-6} m^2/s, i.e. 10^6 cSt = 1 SI unit (unnamed).

Fixed-area Restrictor

Such components may be specifically designed to create a particular resistance at a given flow rate, in which case they may be laminar or turbulent in character.

Laminar

$$\Delta p = KQ \qquad\qquad\qquad\qquad (1.14)$$

Turbulent

$$\Delta p = KQ^2 \qquad\qquad\qquad\qquad (1.15)$$

Fixed-area valves would be included in this classification and eqn 1.15 is expressed more fully as

$$Q = C_d' \; a_o \; \sqrt{\left(\frac{2\Delta p}{\rho}\right)} \qquad\qquad\qquad (1.16)$$

where a_o = orifice area of flow and C_d' = a discharge factor.

Variable-area restrictor

Several hydraulic valves have a variable-area characteristic and the basic flow equation is

$$Q = C_d' \; a_o \sqrt{\left(\frac{2\Delta p}{\rho}\right)}$$

This may be approximated to $Q = k \, a_o \sqrt{\Delta p}$ provided it is

8

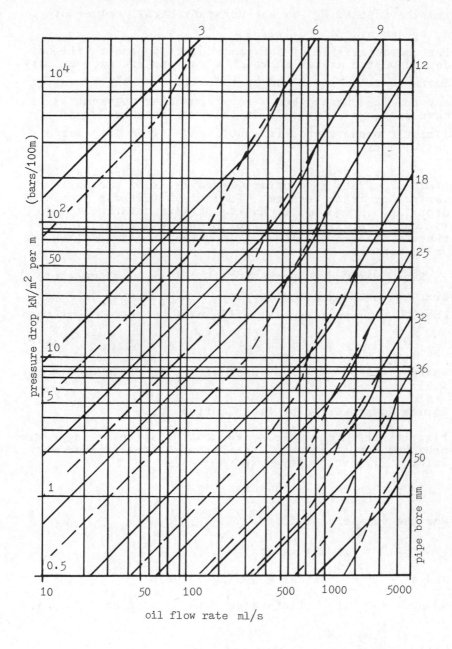

density=870 kg/m^3

viscosity = 50 cSt (full line)
= 25 cSt (dotted line)

Fig.1.7b

9

remembered that C_d' is not constant at all values of Q, Δp and a_o. $(k = C_d'\sqrt{2/\rho})$.

The characteristic of the particular component will now depend on the manner in which a_o and Δp are made to vary. Suitable valve spool and port design can make a_o vary in any required manner but a very common arrangement is to have rectangular ports. In this way the area a_o is directly proportional to spool travel from the closed position (x).

In many systems it is necessary to arrange that the fluid flow rate is constant, regardless of pressure variations in the system. To achieve this the pressure drop (Δp) in eqn 1.16 has to be kept constant. This is done by series or parallel pressure compensation and these systems are discussed in detail in the section on valve control.

With pressure compensation, $C_d'\sqrt{(2\Delta p/\rho)}$ = constant over the working range and so $Q = K'a_o$ and if $a_o = bx$, where b = circumferential width of the control port then

$$Q = Kx \qquad\qquad (1.17)$$

where $K = K'b = C_d'b\sqrt{(2\Delta p/\rho)}$ and Δp is frequently 3 to 4 bar.

Series Compensation of Flow Control Valve

Fig. 1.8 shows a simple speed control system using series pressure compensation of the variable-area orifice valve. For the fixed capacity pump

$$Q_1 = {}_p n_v\, n_p C_p = \text{constant}$$

For the variable orifice

$$Q_2 = K'a_o$$

and $p_2 - p_3$ = constant = spring value m

When $Q_2 < Q_1$ the relief valve must be open and

$$Q_3 = Q_1 - Q_2$$

with p_1 = constant = spring value r

Parallel Compensation of Flow Control Valve

A parallel pressure compensated variable-area orifice

10

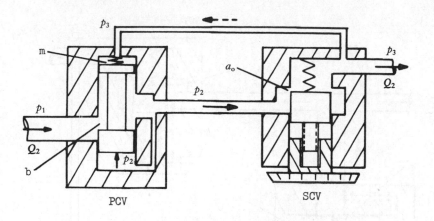

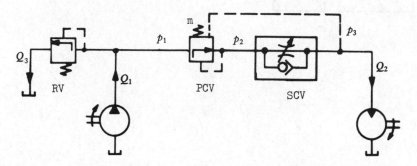

Fig.1.8

valve is illustrated in fig. 1.9. This system of compensation is sometimes called spill-off or by-pass control. The essential feature with this system is the fact that with $Q_2 < Q_1$ the relief valve is not open. The excess oil (Q_3) now passes to tank via the compensating valve. Therefore $Q_3 = Q_1 - Q_2$, $Q_2 = K'a_o$ and $Q_1 = {}_p\eta_v\ n_pC_p$ = constant. Also $p_1 - p_2$ = constant = spring value m.

Efficiency Comparison

Series control

$$\text{Hydraulic system efficiency} = \frac{p_3Q_2}{p_1Q_1}$$

Hence at low speed outputs ($Q_2 \ll Q_1$) and low load ($p_3 \ll p_1$) the system has a very poor efficiency since p_1 is constant at a maximum value.

11

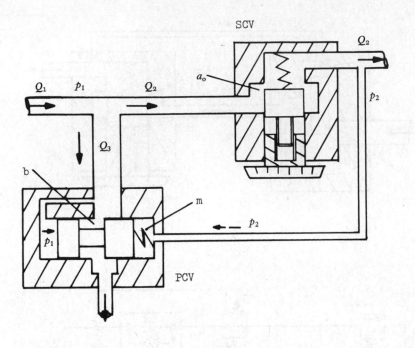

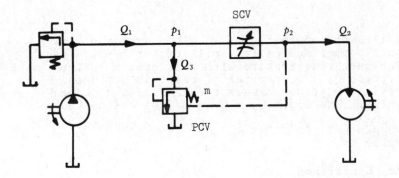

Fig.1.9

Parallel control

$$\text{Hydraulic system efficiency} = \frac{p_2 Q_2}{p_1 Q_1}$$

Here, $p_1 = p_2 + m$, and m is approximately 3 or 4 bar

hence the efficiency depends on the speed only, i.e.
Q_2/Q_1, and is higher than that of a similar series
speed control system.

Pressure-compensated Pump

The variable capacity positive-displacement pump previously discussed will deliver the same fluid flow regardless of load pressure, if leakage is ignored. In
some circuits, particularly where there are actuators
in parallel, it is necessary to reduce the pump output
as the pressure rises. This is known as pressure compensation and is illustrated in a worked example in the
flow-controlled transmissions section.

Basically, the output from the pump is set at the
required maximum value by setting the stroke control.
A spring-loaded actuator then reduces the stroke of the
pump as the pressure rises. This is illustrated in
fig. 1.10. Now

$$Q = \omega_p C_p \, p^n v$$

and $C_p = C_p(\text{max.})$ at $x = 0$

where x = travel of compensating actuator, x_c = pre-
compression of actuator spring and this requires a
pressure $p = p_c$ to overcome it at $x = 0$.

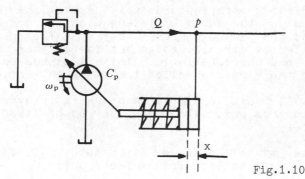

Fig.1.10

$Q = Q(\text{max.})$ at $C_p = C_p(\text{max.})$, i.e. when $x = 0$
$Q = 0$ at $C_p = 0$, i.e. at $x = x(\text{max.})$

Also $p = p(\text{max.})$ at $Q = 0$

and $p = p$ at some value $x = x$

Hence

$$Q = Q(\text{max.}) - \frac{x}{x(\text{max.})} \, Q(\text{max.}) \qquad (1.18)$$

13

From fig. 1.11 it can be seen that eqn 1.18 reduces to

$$Q = Q(\text{max.}) \left[1 - \frac{p - p_c}{p(\text{max.}) - p_c} \right]$$

i.e. $Q = {_p}\eta_v \, \omega_p C_p(\text{max.}) \left[1 - \dfrac{p - p_c}{p(\text{max.}) - p_c} \right]$ \hfill (1.19)

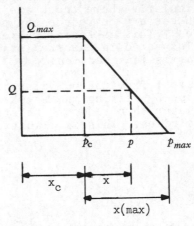

Fig.1.11

Motor

This is a continuous-rotation unit which drives the load. The power output of the motor will depend on its speed of rotation and the pressure drop across it. The speed can be controlled by circuit design but the pressure will depend on the load characteristic. This is discussed in more detail in the flow-controlled transmission section.

Motors are all positive-displacement units in hydro-static transmissions and, like pumps, may be of fixed or variable capacity.

In the steady state the speed of the motor will depend on the oil flow rate to it. For an ideal unit

$$Q_m = C_m n_m \quad (\text{or } C_m \omega_m) \hfill (1.20)$$

where C_m = motor capacity per rev, or radian of shaft rotation

n_m = motor shaft speed, revolutions per unit time

ω_m = motor shaft speed, radians per unit time

Motor leakage is given by an expression similar to that for a pump

14

$$Q_1 = \frac{k \, \Delta p}{\mu} \qquad\qquad (1.21)$$

where Δp is the pressure drop across the motor. To be exact, the leakage will have two components, that due to Δp, i.e. from inlet to outlet, and that due to the pressure p_m, i.e. inlet pressure to motor. However, in most cases the leakage can be expressed in terms of Δp alone. Hence

$$Q_m = C_m \omega_m + \frac{k \, \Delta p}{\mu} \qquad\qquad (1.22)$$

The supply power to the motor is $Q_m \Delta p$ and the volumetric efficiency is

$$_m\eta_v = \frac{Q_m - Q_1}{Q_m} \qquad\qquad (1.23)$$

Hence the power to the rotor of the motor is

$$Q_m \Delta p - Q_1 \Delta p = {}_m\eta_v \, Q_m \Delta p$$

Allowing for mechanical losses between the rotor and the output, then

$$\text{power output} = {}_m\eta_m \, {}_m\eta_v \, Q_m \Delta p$$

where $_m\eta_m$ is the mechanical efficiency of the motor and

$$_m\eta_m = \frac{T_m}{C_m \Delta p} \qquad\qquad (1.24)$$

where T_m = motor shaft torque and C_m is motor capacity per radian of shaft movement

Note

$$_m\eta_o = {}_m\eta_m \, {}_m\eta_v \qquad\qquad (1.25)$$

where $_m\eta_o$ is the overall efficiency of the motor. Again,

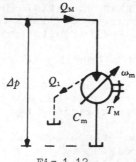

Fig.1.12

15

these are instantaneous efficiencies and are true for only one set of conditions. Fig. 1.12 indicates the steady-state conditions for a motor.

Actuators

These are linear or rotary displacement units of fixed stroke, e.g. piston and cylinder units.

The approach to the steady-state study of these units is by consideration of their 'filling and emptying'. Consider the unit shown in fig. 1.13 where v is the steady piston velocity against a load F and A and a are piston head end and piston rod end areas.

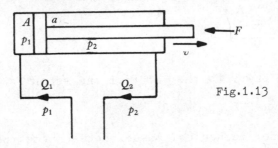

Fig.1.13

$$\text{Flow into cylinder head end} = Q_1 = Av \qquad (1.26)$$
$$\text{Flow out of cylinder rod end} = Q_2 = av \qquad (1.27)$$
$$\text{and} \quad Q_1 = Q_2 + (A - a)v \qquad (1.28)$$

Note: $Q_1 = Q_2$ when $A = a$, i.e. double-rod (or, through-rod) unit.

Also $p_1 A = p_2 a + F$ $\qquad\qquad\qquad\qquad\qquad$ (1.29)

Note: if $A = a$ then

$$p_1 - p_2 = \frac{F}{A} \qquad\qquad\qquad\qquad (1.30)$$

It should also be noted that if $F = 0$ (or is very small) then $p_1 A = p_2 a$. Hence $p_2 = (A/a)p_1$, i.e. the outlet pressure may be greater than the inlet (pump supply) and a relief valve may be required to limit the maximum pressure on this side of the circuit.

Compressibility and Inertia Loading

Steady-state conditions have been considered but two important transient effects must be mentioned. These are compressibility and inertia loading.

When oil is discharged from a pump at high pressure there will be some compression of the oil volume due to the slight compressibility of the oil at high pressure.

16

A volume of oil V_0, at atmospheric pressure will be reduced to a volume V, at a gauge pressure p, according to the relationship

$$V = V_0 \left[1 - \frac{p}{B} \right] \tag{1.31}$$

where B is the bulk modulus of the oil (approximately 17.5 kbar).

The rate of change of volume with pressure is

$$\frac{dv}{dp} = - \frac{V_0}{B} \tag{1.32}$$

Hence the effective flow rate at high pressure is given by

$$Q = \omega_p C_p - \frac{k \, \Delta p}{\mu} - \left[\frac{V_0}{B} \right] \left(\frac{dp}{dt} \right) \tag{1.33}$$

Since Q = volumetric flow rate per unit time.

This effect is illustrated in the examples of the flow controlled transmissions section.

Inertia loading of a system will have an effect during the acceleration periods only. In eqn 1.24 T_m may consist of a steady resisting torque T and an acceleration torque necessary to accelerate a polar moment of inertia J (at the motor shaft). Hence

$$T_m = T + J \left(\frac{d\omega_m}{dt} \right) \tag{1.34}$$

Putting eqn 1.34 into eqn 1.24 gives

$$\Delta p = \frac{T}{m^n{}_m \, Cm} + \left(\frac{J}{m^n{}_m \, Cm} \right) \left(\frac{d\omega_m}{dt} \right) \tag{1.35}$$

Further work on this is covered in the flow- and valve-controlled transmissions sections.

Hydraulic Stiffness

A spring has a stiffness defined as the force required to compress the spring a unit distance. A shaft has a torsional stiffness defined as the torque required to produce unit rotation of the shaft. Due to oil compressibility all hydraulic systems with a trapped volume of oil have a stiffness as shown in the following cases.

Double acting cylinder supplied via a 5/3 directional valve (fig. 1.14).

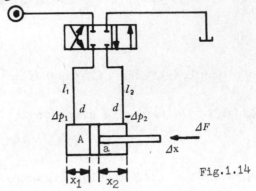

Fig.1.14

If the stiffness is $K = \Delta F/\Delta x$, then when the ports are closed

$$\Delta F = A\,\Delta p_1 + a\,\Delta p_2$$

and $\Delta p_1 = \dfrac{B\,\Delta V_1}{(V_1 + v_1)}$

with $\Delta p_2 = \dfrac{B\,\Delta V_2}{(V_2 + v_2)}$

where $V_1 = Ax_1$ and $V_2 = ax_2$.

Now $v_1 = \pi d^2 l_1/4$ and $v_2 = \pi d^2 l_2/4$, therefore

$$\frac{\Delta F}{\Delta x} = B\left[\frac{A^2}{(V_1 + v_1)} + \frac{a^2}{(V_2 + v_2)}\right]$$

Generally $A = 2a$, hence

$$K = BA^2\left[\frac{1}{(V_1 + v_1)} + \frac{1}{4}\left(\frac{1}{V_2 + v_2}\right)\right]$$

Assuming $v_1 = v_2 = v$, i.e. $l_1 = l_2$ then

$$K = BA^2\left[\frac{1}{(Ax_1 + v)} + \frac{1}{4}\left(\frac{1}{ax_2 + v}\right)\right]$$

$$= BA\left[\left(x_1 + \frac{v}{A}\right)^{-1} + \frac{1}{4}\left(\frac{x_2}{2} + \frac{v}{A}\right)^{-1}\right]$$

hence

$$K = \frac{BA}{2}\left[\frac{(2x_2 + 4b + x_1 + b)}{(x_1 + b)(x_2 + 2b)}\right]$$

18

if $v/A = b$; now

$x_1 + x_2 = X$ (the piston stroke)

hence

$$K = \frac{BA}{2}\left[\frac{2X - x_1 + 5b}{x_1(X + b) - x_1^2 + b(X + 2b)}\right]$$

For min. K, $dK/dx_1 = 0$, hence

$$x_1^2 - x_1(X + b) - b(X + 2b)$$

$$= (2X - x_1 + 5b)(X + b - 2x_1)$$

$$= 2x_1^2 - x_1(5X + 11b) + 2X^2 + 7Xb + 5b^2$$

i.e. $x_1^2 - x_1(4X + 10b) + 2X^2 + 8Xb + 7b^2 = 0$

hence

$$x_1 + (2 - \sqrt{2})X + (5 - 3\sqrt{2})b$$

and $\frac{x_1}{X} = 0.6 + 0.8\frac{b}{X}$

If $b/x \ll 1.0$, i.e. $v/AX \ll 1.0$, i.e. the pipe volume is far less than the max. cylinder volume then $x_1/X = 0.6$ for min. K, hence

$$\text{min. } K = 3.5\frac{BA^2}{V}$$

The reader should establish the following relationships for himself.

Stiffness (K)

$$K = \frac{A^2B}{V}\frac{V_1}{V}\left(1 - \frac{V_1}{V}\right) \quad \text{for fig. 1.15}$$

$$K\text{min.} = \frac{4A^2B}{V}$$

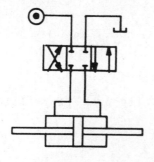

Fig.1.15

A = ram area, V = total ram volume neglecting line volume, V_1 = volume on one side of ram

$$K = A^2B \quad (V_1 + v) \quad \text{for fig. 1.16}$$

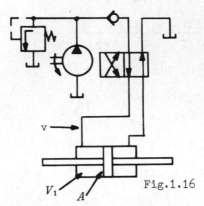

V_1 A Fig.1.16

v = line volume, V_1 = ram volume during stroke, V = total ram volume.

$$K\text{max.} = \frac{A^2B}{v} \qquad K\text{min.} = \frac{A^2B}{V + v}$$

Power pack relief valve normally open (fig. 1.17)

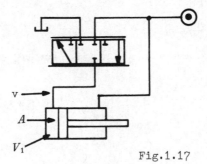

A

V_1

Fig.1.17

$$K = A^2B(V_1 + v)$$

$$K\text{max.} = \frac{A^2B}{v} \qquad K\text{min.} = \frac{A^2B}{V + v}$$

V = max. ram volume at head end.

Relief valves not normally open (fig. 1.18)

$$K = \frac{BC_m^2}{V_0}$$

(note: units of K are torque/radian).

20

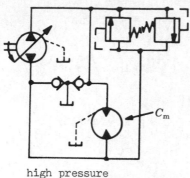

high pressure
line volume $= V_0$

Fig.1.18

Stiffness of a Pneumatic System

The bulk modulus of a gas depends on the type of process involved.

For an isothermal process (constant temperature)

$$pV = mRT$$

where p = absolute pressure   e.g. N/m^2 absolute
      V = volume   e.g. m^3
      m = mass   e.g. kg
      T = absolute temperature e.g. K
      R = gas constant   e.g. Nm/kg K (J/kg K)

If p increases by δp and V decreases by δV then

$$(p + \delta p)(V - \delta V) = mRT$$

thus $\delta p = p \dfrac{\delta V}{V}$

i.e. bulk modulus $= \left(\dfrac{V \delta p}{\delta V}\right)_{\delta \to 0} = p$

For an adiabatic process

$$pV^{\gamma} = \text{constant}$$

where $\gamma = C_p/C_v$
      C_p = specific enthalpy at constant pressure
      C_v = specific enthalpy at constant volume

Hence

$$(p + \delta p)(V - \delta V)^{\gamma} = \text{constant}$$

21

$$(p + \delta p) \, V^{\delta} \left(1 - \frac{\delta V}{V}\right)^{\delta} = \text{constant}$$

$$\delta p = \gamma p \, \frac{\delta V}{V} \quad \text{i.e.} \quad \left(\frac{V \delta p}{\delta V}\right)_{\delta \to 0} = \gamma p$$

i.e. bulk modulus $= \gamma p$

Note $R(\text{air}) = 287$ J/kg K $\gamma(\text{air}) = 1.4$
$R(\text{nitrogen}) = 297$ J/kg K $\gamma(\text{nitrogen}) = 1.6$

System Natural Frequency

Since the compressibility of the oil gives the system a
spring characteristic, then, when combined with a pure
inertia load, there will be a resulting natural frequency for the system.

Natural frequency (rad/s) $\omega_n = \sqrt{(K/M_e)}$ for a linear
system and $\omega_m = \sqrt{(K/J_e)}$ for a rotary system, where M_e =
equivalent mass at actuator output, J_e = equivalent
moment of inertia at motor shaft.

(1) Remember: a mechanical efficiency will have the
effect of giving an apparent increase of inertia. If
the actual moment of inertia of the motor load is J and
the motor mechanical efficiency is η_m then $J_e = J/\eta_m$.

(2) If the motor drives the load via a step down gear
of N:1 (i.e. load runs at 1/N × motor speed), then
$J_e = J/N^2$.

(3) If the gears have a mechanical efficiency of η_G
then $J_e = J/N^2 \eta_G$.

WORKED EXAMPLES

1. Fig. 1.19 shows a hydraulic circuit where the actuator speed is controlled by a meter-in system employing
a series pressure compensated valve. Determine the
power input to the pump for the conditions indicated
(steady state).

If the series compensation is replaced by parallel
compensation, and the load and speed of the actuator
remains unchanged determine the change of overall efficiency of the complete circuit. Note: Q is 1/s, Δp is
bar and pipe losses are neglected.

For valve A, $Q = 0.5\sqrt{\Delta p}$
For valve B, $Q = 0.4\sqrt{\Delta p}$
For valve C, $Q = 0.43\sqrt{\Delta p}$
Piston diameter = 60 mm, rod diameter = 25 mm

For PC valve, $Q = ka$ and $Q = 2.5$ l/s at $k = 8$
For PC valve, $\Delta p = 4$ bar (constant)
For the pump, n(mech.) = 80%, n(vol.) = 92%

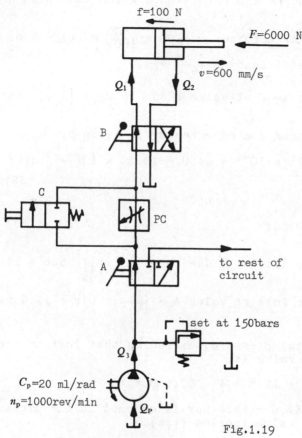

f=100 N

F=6000 N

v=600 mm/s

Q_1

Q_2

B

C

PC

A

to rest of circuit

set at 150bars

Q_3

C_p=20 ml/rad
n_p=1000rev/min

Q_P

Fig.1.19

Series compensation

This means that the pressure drop across the variable
orifice is constant (at 4 bar) and that the excess oil
delivered by the pump flows to tank via the relief valve.

At the actuator

$$Q_1 = \frac{\pi (60)^2}{4} \times 600 \text{ mm}^3/s = 1.7 \ 10^6 \text{ mm}^3/s = 1.7 \text{ l/s}$$

$$Q_2 = 1.7 \left[\frac{60^2 - 25^2}{60^2} \right] = 1.4 \text{ l/s}$$

At the pump

$$Q_p = 2\pi \times \frac{1000}{60} \times 20 \text{ ml/s} = 2.1 \times 10^3 \text{ ml/s} = 2.1 \text{ l/s}$$

23

and $Q_3 = 0.92 \times 2.1 = 1.93$ l/s

Since $Q_3 > Q_1$ then relief valve is open and power input to the pump $= 150 \times 10^5 \times 2.1 \times 10^{-3} \times 1/0.80$ W $= 39.4$ kW

Power output at actuator $= 6100 \times \dfrac{600}{1000}$ watts $= 3.66$ kW

thus system efficiency $= \dfrac{3.66}{39.4} \approx 9\%$

Note pressure loss at valve B due to $Q_2 = \left(\dfrac{1.4}{0.4}\right)^2$ bar $= 12.20$ bar

Pressure at head end of actuator is given by

$$p \tfrac{\pi}{4} (60)^2 \times 10^{-6} = 6100 + 12.20 \times 10^5 \times \tfrac{\pi}{4} (60^2 - 35^2)10^{-6}$$

$$= 6525$$

thus $p = 29.0$ bar

Pressure loss at B due to $Q_1 = \left(\dfrac{1.7}{0.4}\right)^2$ bar $= 18$ bar

Pressure loss at valve A $= \left(\dfrac{1.7}{0.5}\right)^2$ bar $= 11.6$ bar

Therefore total pressure, excluding that lost in pressure compensating valve is

$29 + 18 + 11.6 + 4 = 62.6$ bar

Hence $150 - 62.6 = 87.4$ bar is dropped in the pressure compensating valve (series type).

For a parallel pressure compensating valve the excess oil $(Q - Q_1)$ would bypass at 62.6 bar - 11.6 bar = 51 bar, i.e. this valve would be situated between valve A and the variable orifice (in parallel with the variable orifice). The pump delivery would be at 62.6 bar and so the power to the pump would be

$$62.6 \times 10^5 \times 2.1 \times 10^{-3} \times \frac{1}{0.80} \text{ W} = 16.5 \text{ kW}$$

and the system efficiency would be $3.66/16.5 = 22.2\%$, which is approximately three times as high as before.

2. A valve-controlled system, shown in fig. 1.20, has parallel pressure compensation of the speed-control valve. Determine the conditions for maximum system efficiency and show that this is always less than 50%.

24

For valve C, $Q_1 = m\sqrt{\Delta p_B}$
For valve D, $Q_1 = k\sqrt{\Delta p_A}$ per path

Assuming steady-state conditions with

Q = constant pump delivery
B = constant pressure drop at speed valve
A = effective piston area
F = total load, including friction
v = piston velocity
Δp_A = pressure drop per path of 4/2 valve

Δp_B = pressure drop at non-return valve

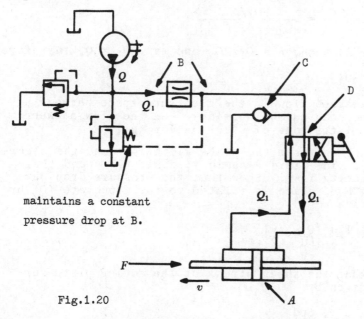

maintains a constant
pressure drop at B.

Fig.1.20

Since the actuator has a double rod with equal areas the
flow into, and out of it, will be the same. The pressure
at the pump outlet will be

$$p = \Delta p_B + \Delta p_A + \frac{F}{A} + \Delta p_A + B$$

and the power at pump outlet will be pQ.

$$pQ = \left[\left(\frac{Q_1}{m}\right)^2 + 2\left(\frac{Q_1}{k}\right)^2 + B + \frac{F}{A} \right] Q$$

The actuator power output is $Fv = FQ_1/A$ therefore

system efficiency $\eta = FQ_1/AQ \left[\left(\frac{Q_1}{m}\right)^2 + 2\left(\frac{Q_1}{k}\right)^2 + B + \frac{F}{A} \right]$

25

Now B = constant, Q = constant and if F is constant, then the maximum efficiency will occur when $d\eta/dQ_1 = 0$

i.e. $FQ\left[\left(\dfrac{Q_1}{m}\right)^2 + 2\left(\dfrac{Q_1}{k}\right)^2 + B + \dfrac{F}{A}\right] = \dfrac{F}{A}Q_1 Q\left[2\dfrac{Q_1}{m^2} + 4\dfrac{Q_1}{k^2}\right]$

thus $B + \dfrac{F}{A} = \left(\dfrac{Q_1}{m}\right)^2 + 2\left(\dfrac{Q_1}{k}\right)^2 - B$

Generally

load pressure $\left(\dfrac{F}{A}\right) = \sum (Q_1^2 \text{ losses}) - \sum (\text{constant losses})$

thus maximum $\eta = \dfrac{FQ_1/A}{2Q[(F/A) + B]}$

If B << F/A, then, $\eta = Q_1/2Q$, and max. $Q_1 = Q$, therefore

$\eta = 50\%$

3. The leakage flow in the clearance space between a piston and body may be considered as the flow between parallel plates, one of which is moving.

For such a system the plate width is b and the clearance is d. The fluid dynamic viscosity is μ and the plate velocity is v. Show that the pressure drop per unit length of plate is related to the flow rate (Q) by the equation

$$\dfrac{\Delta p}{1} = \dfrac{12\mu}{d^2}\left(\dfrac{Q}{bd} - \dfrac{v}{2}\right)$$

Show that the shear stress at the moving plate surface is given by

$$\tau_d = \dfrac{\mu v}{d} \mp \dfrac{\Delta p d}{21}$$

Assumptions

(1) The width b and length 1 are less than the full plate dimensions.
(2) d is small compared with 1 and b, hence the pressure is uniform across the section.
(3) Shear stress is proportional to velocity gradient.
(4) Both plates are horizontal.
(5) Tangential velocity at plates is zero.

Referring to fig. 1.21 let the upper plate be moving with a velocity v as indicated. At the element of thickness δy, distant y from the fixed plate, the viscous shear stress is τ on the lower face and $\tau + \delta\tau$ on the upper face.

$$\pm\, b\delta y\,\Delta p = \tau b1 - (\tau + \delta\tau)b1$$

($\pm\Delta p$ since the pressure drop may be in either direction) hence

$$\pm\,(\Delta p/1)\,\delta y = -\,\delta\tau$$

In the limit $\delta y \to dy$ and $\delta\tau \to d\tau$ hence

$$-\,\tau = \pm\,\Delta py/1 + a$$

where a is a constant.

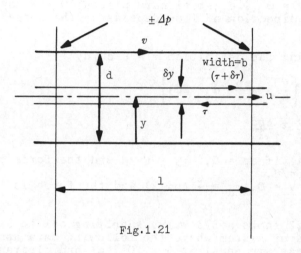

Fig.1.21

The shear stress expression for Newtonian fluids in laminar motion is

$$\tau = \mu\,\frac{du}{dy}$$

where u = velocity at a point distant y from a datum measured at right angles to the flow. Hence

$$-\,\mu\,du = (\pm\,\Delta py/1 + a)\,dy$$

$$-\,u = \pm\,\left[\left(\tfrac{1}{2}\,\Delta py^2/1\right) + ay + b\right]\Big/\mu$$

where b is a constant. But u = 0 at y = 0, i.e. b = 0, and u = v at y = d, i.e. $a = -\,(2\mu 1v \mp \Delta p d^2)/21d$, hence

$$u = \frac{vy}{d} \pm \frac{\Delta p}{2\mu 1}\,(yd - y^2)$$

Now $Q = \displaystyle\int_{0}^{d} ub\,dy$

27

$$= b\int_0^d \frac{vy}{d}\,dy \pm \frac{\Delta p}{2\mu l}\,b\int_0^d (yd - y^2)\,dy$$

thus $Q = bd\left(\dfrac{v}{2} \pm \dfrac{\Delta pd^2}{12\mu l}\right)$

i.e. $\dfrac{\Delta p}{l} = \dfrac{12\mu}{d^2}\left(\dfrac{Q}{bd} - \dfrac{v}{2}\right)$

Note: (i) If $\Delta p = 0$, (i.e. one plate moving over the other inducing viscous flow), then, $Q = vbd/2$.
    (ii) If $v = 0$ (i.e. stationary plates) $Q = \pm\,bd^3\Delta/12$ μl i.e. the direction of flow depends on the sense of Δp.

Again using the definition $\tau = \mu\,du/dy$

$$\tau_d = \mu\left[\frac{v}{d} \pm \frac{\Delta p}{2\mu l}\,(d - 2d)\right]$$

i.e. $\tau_d = \dfrac{\mu v}{d} \mp \dfrac{\Delta pd}{2l}$

Note: (i) If $\Delta p = 0$, $\tau_d = \mu v/d$ and the force on the plate is $\mu lbv/d$.
    (ii) If $v = 0$, $\tau_d = \mp\,\Delta pd/2l$ and the force is $bd\Delta p/2$.

4. Fig. 1.22 shows a 5/3 valve supplying oil to a double actuator system where the following data apply. Actuator areas are equal at 5×10^{-4} m^2 and clearance volume is zero

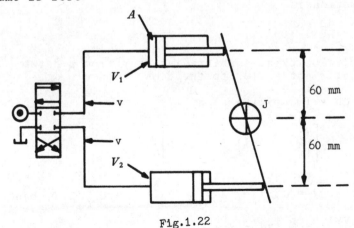

Fig.1.22

Bulk modulus of oil = 15×10^3 bar
Actuator strokes are equal at 30 mm
Moment of inertia of rotating parts at shaft J = 0.05kg m^2

28

Determine the maximum volume of each of the two supply lines between the valve and the actuators if the natural frequency of the system is not to fall below 250 HZ.

$$\omega_n = 2\pi \times 250 = 1570 \text{ rad/s}$$

Now $\omega_n = \sqrt{K/J_e}$ and since J_e is given as 0.05 kg m^2 it is necessary to calculate the minimum stiffness (K) so that ω_n shall not be less than 1570 rad/s

$$K(\text{linear}) = A^2 B \left[\frac{1}{V_1 + v} + \frac{1}{V_2 + v}\right]$$

But $V_1 + V_2 = V = $ constant, hence

$$K = A^2 B \left[\frac{1}{V_1 + v} + \frac{1}{V - V_1 + v}\right]$$

$$= \frac{A^2 B (V + 2v)}{-\left[V_1^2 - V_1 V - v(V + v)\right]}$$

$dK/dV_1 = 0$ therefore $V_1 = V/2$ and

$$K(\text{linear}) = \frac{4A^2 B}{V + 2v} \quad (\text{minimum})$$

Now $K(\text{rotational}) = (0.06)^2 K(\text{linear})$

thus $(\omega_n) = \frac{(0.06)^2}{0.05} \frac{4A^2 B}{V + 2v}$

i.e. $v = 26.9 \text{ ml}$

5. Effect of Entrained Gas in a Liquid

A hydraulic system has an oil volume V_0 at atmospheric pressure p_0. A volume of free gas V_g is mixed with the volume V_0 of oil. If the working pressure of the system is p (absolute) show that for isothermal changes about the working pressure and volume

$$\frac{B(\text{mixture})}{B(\text{oil})} = 1 + \left[\frac{p_0}{p} \frac{V_g}{V_0} - \frac{p - p_0}{B_{oil}}\right] \left[1 + \frac{B_{oil}}{p}\left(\frac{p_0}{p} \frac{V_g}{V_0}\right)\right]$$

At pressure p let oil volume $= V_{oil}$, and gas volume $= V_{gas}$. Consider changes δp, δV_{oil}, δV_{gas} as all positive when increasing. Now

$$B_{oil} = - \frac{V_{oil}}{dV_{oil}} dp$$

$$B_{gas} = p$$

thus total volume of mixture at $p = V_{oil} + V_{gas}$

$$V_{mixture} = \frac{p_0 \, V_g}{p} + V_0 \left[1 - \frac{p - p_0}{B_{oil}} \right]$$

thus
$$\frac{dV_{mixture}}{dp} = - \frac{p_0 \, V_g}{p^2} - \frac{V_0}{B_{oil}}$$

Hence

$$B_{mixture} = - \frac{V_{mixture} \, dp}{dV_{mixture}}$$

$$= - \left[\frac{p_0 \, V_g}{p} + V_0 \left(1 - \frac{p - p_0}{B_{oil}} \right) \right] \bigg/ \left(\frac{p_0 \, V_g}{p^2} + \frac{V_0}{B_{oil}} \right)$$

Thus
$$\frac{B_{mixture}}{B_{oil}} = \left[1 + \frac{p_0}{p} \frac{V_g}{V_0} - \frac{p - p_0}{B_{oil}} \right] \bigg/ \left[1 + \frac{B_{oil}}{p} \left(\frac{p_0}{p} \frac{V_g}{V_0} \right) \right]$$

Normally $B_{oil} \approx 16 \text{ kbar} = 16 \times 10^8 \text{ N/m}^2$ and $p(\text{max.}) = 400\text{bar} = 0.4 \times 10^8 \text{ N/m}^2$

so
$$\frac{B_{mixture}}{B_{oil}} \approx \left[1 + \frac{p_0}{p} \frac{V_g}{V_0} \right] \bigg/ \left[1 + \frac{B_{oil}}{p} \left(\frac{p_0}{p} \frac{V_g}{V_0} \right) \right]$$

Further Example. If the gas compression law is $pV_g^\gamma = $ constant, i.e. adiabatic, show that

$$\frac{B_{mixture}}{B_{oil}} \approx \left[\frac{V_0}{V_g} + \left(\frac{p_0}{p} \right)^{\frac{1}{\gamma}} \right] \bigg/ \left[\frac{V_0}{V_g} + \left(\frac{p_0}{p} \right)^{\frac{1}{\gamma}} \frac{B_{oil}}{\gamma p} \right]$$

Further Examples

1. The speed-control circuit shown in fig. 1.23 has the following data

Component A.
Capacity = 120 ml/rev, shaft speed = 1000 rev/min, volumetric efficiency = 95%

Component B.
Set at 70 bar, assume no pressure over-ride
Component C
Flow equation Q_C = 80 a√Δp where Q_C = ml/min, a = valve flow area, mm², Δp = valve pressure drop, bar, a(max.) = 200 mm²
Component D
Capacity = 160 ml/rev, volumetric efficiency = 95%, mechanical efficiency = 80%, load = constant torque of 60Nm.

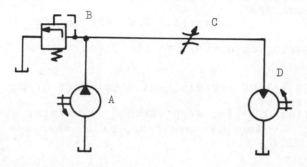

Fig.1.23

Determine the flow through the valve C under these conditions and the maximum speed of D.

(102 l/min 605 rev/min)

2. The series pressure compensated flow control system in fig. 1.24 has a pump (A) with the following details
Capacity = 180 ml/rev, shaft speed = 980 rev/min, volumetric efficiency = 90%, overall efficiency = 75%.
The relief valve is set at 70 bar and may be assumed to have no over-ride.

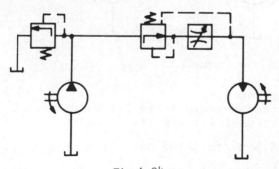

Fig.1.24

Explain the procedure

The compensating valve has a pressure drop of 7 bar when the full pump flow passes through it.
The control orifice has a fixed pressure drop of 3 bar

31

and will just pass the full pump output when fully open.
The hydraulic motor has a capacity of 140 ml/rev, a
volumetric efficiency of 90% and a mechanical efficiency
of 85%.
Determine
  (a) the maximum torque that can be developed at the
motor and the system overall efficiency when the control
valve is fully open
  (b) the system overall efficiency at maximum torque
with 45% maximum power being developed
  (c) the system overall efficiency at 50% maximum torque
and 45% maximum speed

$$[(a)\ 113.5\ \text{N m},\ 49\%\ (b)\ 22\%\ (c)\ 11\%]$$

3. The following data apply to the circuit shown in fig.
1.25.
  Pump
  Shaft speed = 1000 rev/min, volumetric efficiency = 95%
  Variable-area orifice
  Flow equation $Q_c = 100\ a\sqrt{\Delta p}$ where Q_c = ml/min, a = flow
area mm^2, Δp = valve pressure drop, bar. When a = 300
mm^2, $Q_m = 0.72\ Q_p$

  Motor
  Capacity = 80 ml/rev, shaft speed = 860 rev/min, volu-
metric efficiency = 93%, overall efficiency = 68%
  Relief valve
  Q_B = K × (over-ride pressure) × $\sqrt{\text{(system pressure)}}$

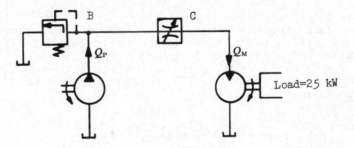

Fig.1.25

Determine
  (a) The pump capacity
  (b) the system pressure at relief valve
  (c) the valve of K for the relief valve.

$$(108.5\ \text{ml/rev},\ 306\ \text{bar},\ 21\ \text{ml min}^{-1}\ \text{bar}^{-3/2})$$

4. The circuit shown in fig. 1.26 is for a hydraulic
hoist where the maximum load to be raised is 13.4 kN at
0.6 m/s.
The 'empty' lowering load of the hoist is 450 N. Neglect
pulley and ram friction. The ram piston diameter is 100

mm and the rod area is half the piston area. The pump
overall efficiency is 85%. The flow rate/pressure drop
relationship for the ports of valve H using flow paths,
A → B, C → D, A → C, B → D, A → D, is Q 1/min = 140√Δp
bar and for valves J and G is Q 1/min = 180√Δp bar.

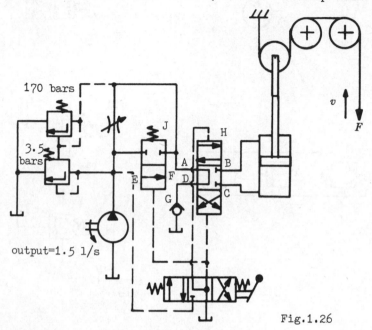

Fig.1.26

Determine for the steady-state
  (a) The power being dissipated across valve A while
the maximum load is being raised at 0.6 m/s.
  (b) The overall efficiency of the system under these
conditions.
  (c) The maximum lowering speed of the hoist when
empty.
Assume that the metering valve is being totally by-
passed during (c).
        [(a) 2.51 kW (b) 60% (c) 0.38 m/s]

5. The hydraulic components shown in fig. 1.27 may be
assembled to form a meter-in speed-control system using
series or parallel pressure compensation. The pressure
drop per flow path in valve A is given by Δp (bar) =
$3 \times 10^{-5} Q^2$ where Q = ml/s.
Sketch both these arrangements and show that for a constant pressure drop across the motor at all speeds (say
5.5 bar) the parallel system is the more efficient.

6. A power transmission system employs 25 m of 20 mm
bore tube with oil of density 870 kg/m^3 and kinematic
viscosity of 40 cSt. For a flow rate of 0.314 1/s
determine the pressure loss in the tube. If the flow

33

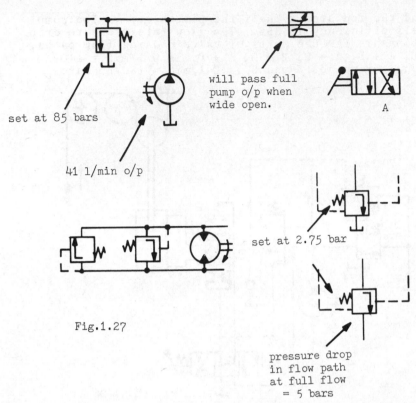

set at 85 bars

will pass full
pump o/p when
wide open.

A

41 1/min o/p

set at 2.75 bar

Fig.1.27

pressure drop
in flow path
at full flow
= 5 bars

rate is increased to 2.512 1/s determine the new pressure loss.
Assume that, for turbulent flow, the friction coefficient $f = 0.08(Re)^{-0.25}$ in the expression $\Delta p = 4fl\rho v^2/2d$.

(0.694 bar; 11.8 bar)

7. Oil of specific gravity 0.8 and kinematic viscosity 186 mm^2 s^{-1} flows through a 300 m smooth pipe connecting two tanks where the surface level difference is 16 × 10^{-2} m. Ignoring losses other than pipe friction, determine the size of pipe necessary for a flow rate of 89 1/s.

(0.6 m diameter)

8. Laminar flow of a fluid, of dynamic viscosity μ, through a straight circular pipe of length 1 and diameter d, results in a pressure drop Δp in the direction of flow. Show that, for a mean velocity of flow v the pressure drop is given by $\Delta p = 32\mu lv/d^2$.

9. Fig. 1.28 shows the arrangement of a flapper nozzle measuring device. The flapper (A) moves relative to the nozzle (B) and gives a flow area proportional to

34

the gap distance x. The flow equation for the nozzle
is $Q_2 = K_2 x \sqrt{\Delta p_2}$ where K_2 is a constant and Δp_2 is the
pressure drop across the nozzle. The flow equation for
the restriction is $Q_1 = K_1 \sqrt{\Delta p}$ where K_1 is a constant

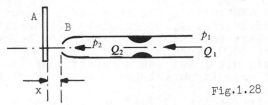

Fig.1.28

and Δp_1 the pressure drop across the restriction. The
fluid is incompressible; p_1 is a constant supply pressure. Plot a graph showing the variation of p_2/p_1 with
$K_2 x/K_1$ and show, by analysis, that maximum sensitivity
occurs when $p_2/p_1 = 3/4$. (Sensitivity is defined as the
change of p_2 for a given change of x.)

10. The valve shown in fig. 1.29 is a combined series
pressure-compensating valve and a variable-area orifice.
The flow equation for the orifice is of the form Q =
$C a\sqrt{\Delta p}$ where Q = 1/min, a = valve flow area (mm²) and
Δp = orifice pressure drop (bar). Under test it was
found that at a maximum flow area of 25 mm² the flow
was 50 1/min with p_1 = 79 bar and p_2 = 72.75 bar. The
compensating spring has a stiffness of 50 N/mm and an
initial compression of 3 mm when the compensating valve
is at full-flow area. The compensating piston area is
400 mm² and the displacement x = 8 mm at full-flow area.

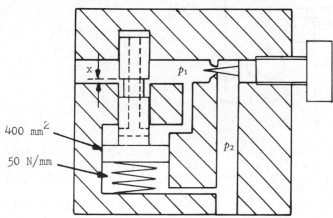

Fig.1.29

Determine the value of x for the test flow conditions
stated above. If the variable orifice-area remains
unchanged but the associated pressure drop increases by
2% determine the new flow rate through the valve and the

35

value of the coefficient C. Determine also the new
value of x.

$$(6 \text{ mm, } 50.5 \text{ 1/min, } 0.8, \text{ } 5.9 \text{ mm})$$

11. A method of controlling the speed of a linear actuator is indicated in fig. 1.30. Explain the operation
of the circuit. The following data apply to the circuit.
Both pumps run at 1000 rev/min
Supply pump at circuit inlet has a capacity of 150 ml/rev
Metering pump at circuit outlet has a capacity of 40 ml/
rev
Actuator piston area = 5×10^3 mm^2 and rod area = $2.5 \times$
10^3 mm^2
The load on the actuator (F) = 20 kN and is effective
when cam A has completely cleared valve B. The flow
equation for valve C is $Q = 40\sqrt{\Delta p}$ for each flow path,
and for B $Q = 17.5\sqrt{\Delta p}$ where $Q = $ 1/min, $\Delta p = $ flow path
pressure drop (bar). The main relief valve is set at
54 bar and has no over-ride.

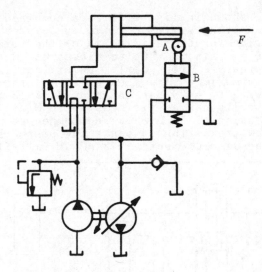

Fig.1.30

Determine the actuator rod-end pressure when loaded and
the power being dissipated at the relief valve. Calculate the fast-forward, slow-forward and fast-return speeds
of the actuator piston.

$$(20 \text{ bar, } 6.3 \text{ kW, } 30 \text{ m/min, } 16 \text{ m/min, } 60 \text{ m/min})$$

12. The following data apply to the system of fig. 1.31
Input pump capacity = C_1 per rev
Metering pump capacity = C_2 per rev where $C_2 < C_1$
Common shaft speed = N rev/unit time
Flow equation per flow path of each valve: $Q = k\sqrt{\Delta p}$
Relief valve set at p_R with no over-ride

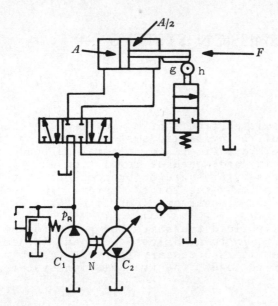

Fig.1.31

The load F is active when the cam g is clear of valve h.

Under these conditions show that the maximum efficiency of the loaded circuit may be expressed as

$$\eta = \frac{F}{Ap_R} \left[\frac{3}{4} \frac{C_1}{C_2} - 1 + \frac{F}{Ap_R} \right]^{-1}$$

2   TRANSMISSION SYSTEMS

INTRODUCTION

Power is transmitted from a prime mover or an electric
motor by use of the hydraulic system. The pump (i)
circulates the hydraulic fluid at a rate dependent on
the pump capacity and speed of revolution and (ii) provides the necessary pressure to overcome the total load
reflected on to it by the loaded system. The pump is
driven by the prime mover or electric motor.

The hydraulic fluid transmits the power to the hydraulic motor which, in turn, drives the load. Hence in the
complete system it is necessary to study the characteristics of the prime mover, the transmission system and the
load.

THE PRIME MOVER

This will be the source of power that is to be transmitted and may take the form of

(a) a diesel engine
(b) a petrol engine
(c) a steam turbine
(d) a gas turbine
(e) an electric motor

In the initial work it will be assumed that the prime
mover maintains a constant pump shaft speed and provides
the power required. Eventually it will be necessary to
allow for the dynamic response of the prime mover in
order to study the complete system characteristics.

THE TRANSMISSION SYSTEM

This will consist of the pump, motor and valves employed
to control the system i.e. control fluid flow rate, fluid
pressure and direction of flow. In electro-hydraulic
systems, electronic and electrical components will form
part of the control section of the transmission system
and will need to be included in the analysis.
(a) If both the pump and the motor are fixed-capacity
units then fluid control is achieved by the use of valves
alone.
(b) If the pump or the motor is a variable-capacity
unit then variation of its capacity will have a control
effect on the system. The manner employed to vary the
capacity of the element will need to be chosen to suit
the requirements of the system.

(c) If the demand for power transmission is not continuous then an analysis of the exact demand, in magnitude and frequency, will need to be undertaken in order to design a suitable system.

These three types of system will be considered in separate chapters as (a) valve control systems (b) flow control systems and (c) accumulator systems.

THE LOAD

The load on a transmission system may be difficult to define in simple mathematical terms but, in order to facilitate an understanding of the subject, only simply loading systems will be considered

(a) constant torque and constant force
(b) constant power
(c) pure inertia
(d) inertia with viscous friction

Expressed mathematically these are
(a) motor torque T_m = T, ram force F_m = F
(b) motor power output P_m = $T_m \omega_m$ = P,

ram power P_m = $F_m v_m$ = P
(c) motor torque = T_m = J $d\omega_m/dt$, ram force =

F_m = M dv_m/dt
(d) motor torque = T_m = J $\dfrac{d\omega_m}{dt}$ + $f\omega_m$

ram force = F_m = M $\dfrac{dv_m}{dt}$ + fv_m

Here it has been assumed that mass M and moment of inertia J are constant. Also: ω_m = angular speed, v_m = linear speed of motor, f = viscous torque/unit angular velocity or viscous force/unit linear velocity.

FLOW CONTROL SYSTEMS

Component Efficiencies

For the pump

Q	= delivered flow rate
Q_1	= leakage flow rate
$\omega_p C_p$	= pump throughput
ω_p	= shaft speed
Δp_p	= pressure rise
T_p	= drive torque to pump

C_p = pump capacity/radian

From fig. 2.1

$$Q = \omega_p C_p - Q_1$$

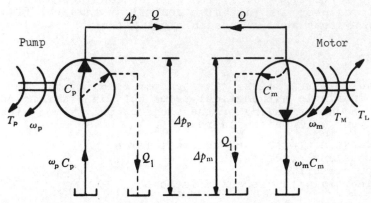

Fig.2.1

Volumetric efficiency $\eta_{vp} = \dfrac{Q}{\omega_p C_p}$

Mechanical efficiency $\eta_{mp} = \dfrac{\omega_p C_p \, \Delta p_p}{\omega_p T_p} = \dfrac{C_p \, \Delta p_p}{T_p}$

Overall efficiency $\eta_{op} = \dfrac{Q \, \Delta p_p}{\omega_p T_p}$

Also $\eta_{op} = \eta_{vp} \times \eta_{mp}$

For the motor

Q = oil supply flow rate ω_m = shaft angular speed
Q_1 = leakage T_L = load torque on motor
$\omega_m C_m$ = motor throughput C_m = motor capacity/radian
Δp_m = pressure drop T_m = motor torque

From Fig. 2.1

$$Q = \omega_m C_p + Q_1$$

Volumetric efficiency $\eta_{vm} = \dfrac{\omega_m C_m}{Q}$

Mechanical efficiency $\eta_{mm} = \dfrac{\omega_m T_m}{\omega_m C_m \, \Delta p_m} = \dfrac{T_m}{C_m \, \Delta p_m}$

40

Overall efficiency $\eta_{om} = \dfrac{\omega_m T_m}{Q \Delta p_m}$

Also $\eta_{om} = \eta_{vm} \times \eta_{mm}$

Leakage

The leakage path will normally be of small cross-sectional area compared with the total wetted area and the flow velocity will be small. Because of these facts it is reasonable to assume laminar flow hence

$$Q_1 = \frac{k}{\mu} \Delta p$$

where μ = fluid dynamic viscosity, therefore

$$Q_1 = \lambda \Delta p$$

where λ is the component leakage coefficient for a particular fluid at a particular temperature. Throughout the analysis to follow λ will be assumed to be a constant and will be called the leakage coefficient even though it has dimensions.

Compressibility

All hydraulic fluids are compressible but liquids need to be subjected to considerable pressure rise before any appreciable reduction in volume results.

For the analysis to follow a linear relationship will be assumed between volume change and applied pressure.

$$V = V_0 \left(1 - \frac{\Delta p}{B}\right)$$

where V = volume resulting from applying a pressure rise Δp to a volume V_0 of liquid of bulk modulus B.

Since the liquid will have to be contained in a vessel then a modified value of B will be employed that will allow for the elasticity of the constraining boundary.

$$\frac{1}{B'} = \frac{1}{B} + \frac{1}{B_e}$$

where B_e = effective bulk modulus of constraining boundary and B' = modified fluid bulk modulus. Therefore

$$V = V_0 \left(1 - \frac{\Delta p}{B'}\right)$$

Hence there will be a reduction in the value of Q delivered by the pump given by

41

$$Q = \omega_p C_p - Q_1 - \frac{V_0}{B'} \frac{d(\Delta p_p)}{dt}$$

Linearity

All the relationships established so far are linear ones and for simplicity these linear relationships will be assumed to exist over the whole working range.

Transmission Circuits

In figs 2.2 and 2.3 V_0 = total volume on high pressure side of the circuit, including those parts of the pump and motor exposed to the line, when $\Delta p = 0$.

Fig. 2.2 shows a system where the motor can rotate in one direction only unless a reversing valve is introduced between the pump and motor.

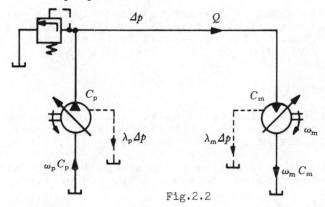

Fig.2.2

Fig. 2.3 shows a reversible system where a boost pump makes up the leakage loss and maintains a minimum pressure on the low pressure side of the system.

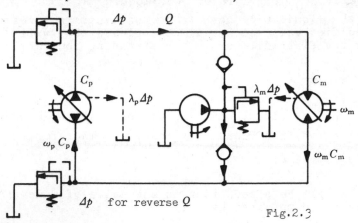

Fig.2.3

42

Speed

For both circuits

$$Q = \omega_p C_p - \lambda \Delta p - \frac{V_0}{B'} \frac{d(\Delta p)}{dt} = \omega_m C_m + \lambda_m \Delta p$$

$$\text{motor speed } \omega_m = \omega_p \frac{C_p}{C_m} - \frac{\Delta p}{C_m}(\lambda_p + \lambda_m)$$

$$- \frac{V_0}{C_m B'} \frac{d(\Delta p)}{dt} \qquad (2.1)$$

Neglecting leakage and compressibility
(a) if the motor capacity is fixed and the pump speed is constant then $\omega_m = kC_p$ (see fig. 2.4a) and the maximum motor speed is set by maximum value of C_p. For $\omega_m(\text{max.}) > \omega_p$ then $C_p(\text{max.}) > C_m$.

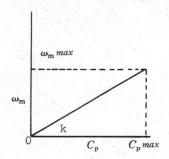 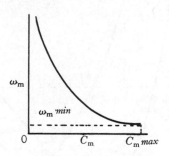

Fig.2.4a   Fig.2.4b

(b) if the pump capacity is fixed and the pump speed is constant then $\omega_m = k/C_m$ (see fig. 2.4b) and the minimum motor speed is set by the maximum value of C_m. The maximum motor speed will occur at $C_m = 0$ i.e. $\omega_m = \infty$. In practice the system pressure will rise due to friction as ω_m increases and eventually the relief valve will open, but this could still be at a dangerously high value of ω_m so a mechanical limit to set the minimum value of motor capacity may be necessary.
(c) if both pump capacity and motor capacity are variable then the motor speed at any instant will depend on the relative value of $C_p/C_p(\text{max.})$ and $C_m/C_m(\text{max.})$.

Ignoring leakage and compressibility, the speed ω_m is given by

43

$$\omega_m = \omega_p \frac{C_p}{C_m} \quad \text{i.e.} \quad \frac{\omega_m}{\omega_p} = \frac{C_p}{C_m}$$

(i) Now putting $C_m = C_m$(max.) and keeping it constant at this value then in fig. 2.5 line OA represents the variation of ω_m with increasing C_p up to $C_p = C_p$(max.).

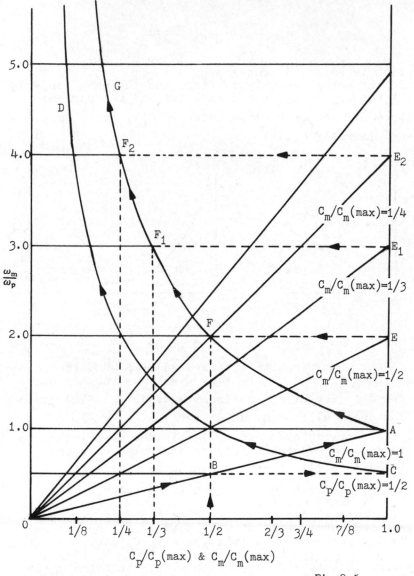

Fig.2.5

If C_p is now maintained at C_p(max.) and C_m is gradually reduced, the variation of ω_m with C_m will follow the curve AG.

(ii) If C_m is set at $\frac{1}{2}C_m$(max.) and C_p gradually increased from 0 to C_p(max.) then line OE will show the variation of motor speed ω_m with pump capacity C_p up to $C_p = C_p$(max.). The variation of ω_m with decreasing C_m (C_p fixed at C_p(max.)) will be given by curve FG.

(iii) If motor capacity C_m is set at C_m(max.) and the pump capacity C_p is increased from 0 to $\frac{1}{2}C_p$(max.) then line OB will indicate the variation of motor speed ω_m. If C_p is now fixed at $\frac{1}{2}C_p$(max.) and C_m gradually reduced from C_m(max.) then curve CD will show the variation of motor speed ω_m.

Load

Consider a hydraulic motor, as shown in fig. 2.6, subjected to various loadings. In each case the relationship between the load and the system pressure will be developed and substituted in the speed equation 2.1 to give the equation governing the response of the system.

Based on the definitions already stated

$$\text{mechanical efficiency } \eta_{mm} = \frac{T_m}{\Delta p C_m}$$

$$\text{overall efficiency } \eta_{om} = \frac{\omega_m T_m}{\Delta p \, Q}$$

(a) Constant torque load

Motor torque $T_m = T$

therefore

$$\Delta p = \frac{T}{\eta_{mm} C_m} \qquad\qquad (2.2)$$

If C_m is constant then Δp is as constant as the mechanical efficiency will permit. Substituting eqn 2.2 into eqn 2.1 gives

$$\omega_m = \omega_p \frac{C_p}{C_m} - \frac{T(\lambda_p + \lambda_m)}{C_m{}^2 \, \eta_{mm}}$$

45

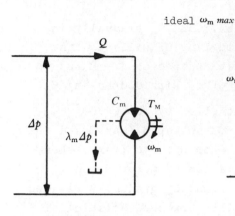

speed slip due to leakage
and mechanical losses

ideal ω_m *max*

ω_m ideal

actual

0   C_p C_p *max*

small C_p providing leakage
only at zero motor speed

Fig.2.6

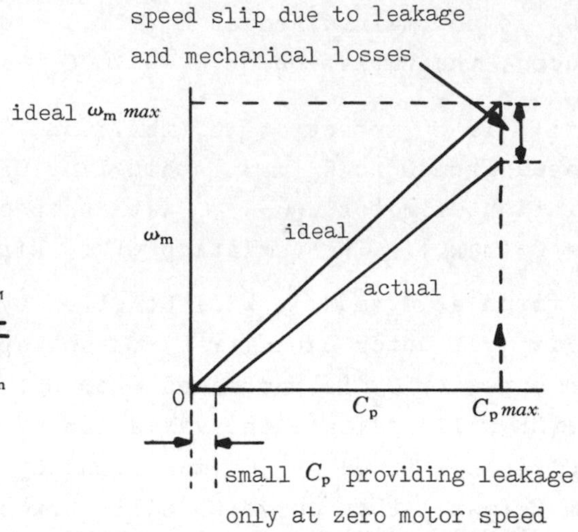

Fig.2.7

Hence the greater the torque the greater will be the speed 'slip' due to leakage.

A decreasing mechanical efficiency with increasing speed will also produce an increased speed 'slip' as indicated in fig. 2.7. Such an arrangement is often called a 'constant torque transmission system' since $\Delta p \simeq$ constant for T_m = constant.

Reconsidering eqn 2.2, $\Delta p = T/(\eta_{mm} C_m)$, if now C_m is variable (and C_p fixed) then eqn 2.2 may be rewritten as

$$\Delta p = \frac{\omega_m T}{\eta_{mm} C_m \omega_m} = \frac{\omega_m T}{\eta_{mm} \eta_{vm} Q}$$

therefore

$$\Delta p = \omega_m \frac{T}{\eta_{om} Q} \qquad (2.3)$$

i.e. $\Delta p \simeq \omega_m \times$ constant

Substituting eqn 2.3 into eqn 2.1 gives

$$\omega_m = \omega_p \frac{C_p}{C_m} - \omega_m \frac{T}{\eta_{om} Q C_m}(\lambda_p + \lambda_m) - \frac{V_0}{C_m B'} \frac{T}{\eta_{om} Q} \frac{d\omega_m}{dt}$$

therefore

$$\omega_m C_m + \frac{\omega_m T}{\eta_{om} Q}(\lambda_p + \lambda_m) + \frac{V_0 T}{B' \eta_{om} Q} \frac{d\omega_m}{dt} = \omega_p C_p \qquad (2.4)$$

46

Eqn 2.4 is of the form

$$\omega_m(s) \, C_m(s) + \frac{T(\lambda_p + \lambda_m)}{\eta_{om} Q} \, \omega_m(s) +$$

$$\frac{V_0 T}{B'\eta_{om}Q} \left[s\omega_m(s) - \omega_m(\min.) \right] = \omega_p C_p$$

if $\omega_m = \omega_m(\min.)$ at $t = 0$ and $s = $ Laplace operator. The steady-state solution in the time domain is

$$\omega_m = \frac{\omega_p C_p}{C_m + \frac{T(\lambda_p + \lambda_m)}{\eta_{om}Q}}$$

It can be seen that the leakage introduces a speed 'slip' which increases with increasing values of torque T.

(b) Constant power load

Motor power $P_m = \omega_m T_m = P$

Overall efficiency $\eta_{om} = \frac{P}{\Delta p \, Q}$

therefore

$$\Delta p = \frac{P}{\eta_{om}Q} \qquad\qquad (2.5)$$

If C_p is fixed and C_m variable then Q is approximately constant, thus $\Delta p \simeq$ constant, assuming $\eta_{om} \simeq$ constant.

Substituting eqn 2.5 into eqn 2.1 gives

$$\omega_m = \omega_p \frac{C_p}{C_m} - \frac{P}{\eta_{om}QC_m} (\lambda_p + \lambda_m)$$

$$= \frac{1}{C_m} \, \omega_p C_p - \frac{P(\lambda_p + \lambda_m)}{\eta_{om}Q}$$

Hence the greater the power being transmitted the greater will be the speed 'slip'. Such an arrangement is often called a 'constant power transmission system' since $\Delta p \simeq$ constant.

Reconsidering eqn 2.5, $\Delta p = P/(\eta_{om}Q)$, if now C_p is variable and C_m fixed then eqn 2.1 may be rewritten as

$$\omega_m = \omega_p \frac{C_p}{C_m} - \frac{P}{\eta_{om}QC_m} (\lambda_m + \lambda_p) - \frac{V_0}{C_m B'} \frac{d}{dt} (P/\eta_{om}Q)$$

Hence as C_p is increased from zero, ω_m is also increased but since Q is increasing the effect of the leakage term is reduced. The compressibility term

$$- \frac{V_O}{C_m B'} \frac{P}{\eta_{om}} \frac{d}{dt} \left(\frac{1}{Q}\right)$$

can be expressed as

$$+ \frac{V_O}{C_m B'} \frac{P}{\eta_{om} Q^2} \frac{dQ}{dt}$$

Hence an increasing Q i.e. $dQ/dt = +$ ve will give a slightly increased value of ω_m since this term is now positive (use eqn 2.5 to explain this). Since Q appears in the denominator the total effect of this term is not likely to be very great.

(c) Pure inertia load

$$T_m = J \frac{d\omega_m}{dt}$$

and $\eta_{mm} = \dfrac{T_m}{\Delta p \; C_m}$

therefore

$$\Delta p = \frac{J}{\eta_{mm} C_m} \frac{d\omega_m}{dt} \tag{2.6}$$

Substituting eqn 2.6 into eqn 2.1 gives

$$\omega_m = \omega_p \frac{C_p}{C_m} - \frac{J(\lambda_p + \lambda_m)}{\eta_{mm} C_m{}^2} \frac{d\omega_m}{dt} - \frac{V_O}{\eta_{mm} B' C_m{}^2} J \frac{d^2\omega_m}{dt^2} \tag{2.7}$$

i.e. $\dfrac{d^2\omega_m}{dt^2} + \dfrac{B'(\lambda_p + \lambda_m)}{V_O} \dfrac{d\omega_m}{dt} + \dfrac{\eta_m B' C_m{}^2}{JV_O} \omega_m$

$$= \frac{\eta_m B' C_m{}^2}{JV_O} \omega_p \frac{C_p}{C_m} \tag{2.8}$$

(note $(\eta_m B' C_m{}^2/JV_O)^{\frac{1}{2}}$ = system natural frequency ω_n)

Consider now the condition that compressibility can be neglected. Eqn 2.7 can be written as

48

$$\frac{J(\lambda_p + \lambda_m)}{\eta_{mm}C_m{}^2} \frac{d\omega_m}{dt} + \omega_m = \omega_p \frac{C_p}{C_m} \qquad (2.9)$$

Hence for a fixed-capacity motor/variable-capacity pump system eqn 2.9 may be rewritten as

$$(\tau D + 1)\omega_m = \omega_p \frac{C_p}{C_m} \qquad (2.10)$$

where the time constant

$$\tau = \frac{J(\lambda_p + \lambda_m)}{\eta_{mm}C_m{}^2}$$

and $D = d/dt$. From this it can be seen that large moment of inertia J or large leakage rates produce a large time constant τ and associated slow response.

Eqn 2.8 may be rewritten as

$$(D^2 + 2\zeta\omega_n D + \omega_n{}^2)\omega_m = \omega_n{}^2 \omega_p \frac{C_p}{C_m} \qquad (2.11)$$

where the damping rate $\zeta = \dfrac{B'(\lambda_p + \lambda_m)}{2V_0} \dfrac{1}{\omega_n} = \dfrac{\tau\omega_n}{2}$

and the natural frequency

$$\omega_n = C_m \sqrt{\left(\frac{\eta_{mm}B'}{JV_0}\right)} = \sqrt{\left(\frac{K\eta_{mm}}{J}\right)} \qquad (2.12)$$

where K = stiffness of system.

From this it can be seen that

$$\zeta = \frac{(\lambda_p + \lambda_m)}{2C_m} \sqrt{\left(\frac{JB'}{\eta_{mm}V_0}\right)} \qquad (2.13)$$
$$= \frac{\tau\omega_n}{2}$$

i.e. increased leakage increases ζ and decreases the oscillation during the transient response. Also increased inertia J increases ζ, which is perhaps the opposite to the effect that would normally be expected by increasing a system's inertia.
It should be remembered that the actual frequency of damped vibration ω_d is given by

$$\omega_d = \omega_n \sqrt{(1 - \zeta^2)} \qquad (2.14)$$

49

Further analysis is given in the worked examples included at the end of this chapter.

(d)   Inertia load with viscous damping on output shaft

$$T_m = J \frac{d\omega_m}{dt} + f\omega_m$$

where f = resisting torque per unit angular velocity. Also

$$\eta_{mm} = \frac{T_m}{\Delta p \, C_m}$$

therefore

$$\Delta p = \frac{J}{\eta_{mm} C_m} \frac{d\omega_m}{dt} + \frac{f}{\eta_{mm} C_m} \omega_m \tag{2.15}$$

Substituting eqn 2.15 into eqn 2.1 gives

$$\omega_m = \omega_p \frac{C_p}{C_m} - \frac{J(\lambda_p + \lambda_m)}{\eta_{mm} C_m{}^2} \frac{d\omega_m}{dt} - \frac{f(\lambda_p + \lambda_m)}{\eta_{mm} C_m{}^2} \omega_m$$

$$- \frac{V_O}{\eta_{mm} B' C_m{}^2} J \frac{d^2\omega_m}{dt^2} - \frac{V_O f}{\eta_{mm} B' C_m{}^2} \frac{d\omega_m}{dt}$$

i.e.
$$\frac{d^2\omega_m}{dt^2} + \left[\frac{B'(\lambda_p + \lambda_m)}{V_O} + \frac{f}{J} \right] \frac{d\omega_m}{dt} + \left[\frac{\eta_{mm} B' C_m{}^2}{JV_O} \right.$$

$$\left. + \frac{f(\lambda_p + \lambda_m)B'}{JV_O} \right] \omega_m = \frac{\eta_{mm} B' C_m{}^2}{JV_O} \omega_p \frac{C_p}{C_m} \tag{2.16}$$

and if compressibility can be neglected

$$\frac{J(\lambda_p + \lambda_m)}{\eta_{mm} C_m{}^2 + f(\lambda_p + \lambda_m)} \frac{d\omega_m}{dt} + \omega_m =$$

$$\frac{\eta_{mm} C_m{}^2}{\eta_{mm} C_m{}^2 + f(\lambda_p + \lambda_m)} \omega_p \frac{C_p}{C_m} \tag{2.17}$$

From eqn 2.16 it can be seen that the new natural frequency ω_n is given by

$$\omega_n = C_m \sqrt{\left[\frac{B'\eta_m}{JV_O} \left\{ 1 + \frac{f(\lambda_p + \lambda_m)}{\eta_{mm} C_m{}^2} \right\} \right]} \tag{2.18}$$

(compare this with eqn 2.12)

Also $\zeta = \dfrac{(\lambda_p + \lambda_m)}{2C_m} \sqrt{\left[\dfrac{JB'}{\eta_{mm}V_0} \dfrac{1}{1 + \dfrac{f(\lambda_p + \lambda_m)}{\eta_{mm}C_m^2}}\right]}$

$+ \dfrac{f}{2C_m} \sqrt{\left[\dfrac{V_0}{JB'\eta_{mm}} \dfrac{1}{1 + \dfrac{f(\lambda_p + \lambda_m)}{\eta_{mm}C_m^2}}\right]}$ (2.19)

Eqn 2.19 indicates that the new damping ratio consists of two parts, one being mainly dependent on leakage and this part increases in value as the inertia J increases. The other part is due mainly to friction and this part decreases as J increases. (Compare eqn 2.19 with eqn 2.13.)

Eqn 2.17 could be expressed as

$$(\tau'D + 1) \omega_m = \dfrac{1}{1 + \dfrac{f(\lambda_p + \lambda_m)}{\eta_{mm}C_m^2}} \omega_p \dfrac{C_p}{C_m}$$ (2.20)

where the time constant

$$\tau' = \dfrac{J(\lambda_p + \lambda_m)}{\eta_{mm}C_m^2} \dfrac{1}{1 + \dfrac{f(\lambda_p + \lambda_m)}{\eta_{mm}C_m^2}}$$

τ' of eqn 2.20 should be compared with τ of eqn 2.10. Further analysis is included in the worked examples.

The reader should establish that eqn 2.16 may be written as

$$\dfrac{1}{\omega_n^2} s^2 + \left(\tau + \dfrac{f}{J} \dfrac{1}{\omega_n^2}\right) s + \left(1 + \dfrac{f}{J}\tau\right) \omega_m = \omega_p \dfrac{C_p}{C_m}$$

where ω_n and τ are as previously defined and s is the Laplace operator.

WORKED EXAMPLES

1. A hydrostatic transmission system consists of a variable-capacity pump and a fixed-capacity motor. The maximum flow between the pump and motor is 500 ml/s and the

motor capacity is 25 ml/rev. The maximum allowable circuit pressure is 70 bar and all efficiencies are to be taken as 100%. Determine the maximum power, maximum speed and maximum torque available at the motor output shaft.

If a constant power output of 2kW is to be developed determine (a) the minimum speed at which this power can be developed and (b) the torque available when the motor is running at maximum speed.

If, below the speed obtained from (a) above, the torque is maintained constant, determine the speed at which 20% of the system's maximum power output is being developed.

Maximum power is developed at maximum pressure and flow rate, therefore

$$P(max.) = (\Delta p \; Q)max. = 7 \times 10^6 \times 500 \times 10^{-6} = 3500 \text{ W}$$

$$\text{Maximum motor speed} = \frac{500}{25} = 20 \text{ rev/s}$$

for 100% volumetric efficiency. For 100% mechanical efficiency

$$\eta_{mm} = \frac{T_m}{\Delta p C_m}$$

Max. torque $= 70 \times 10^5 \times 25 \times \frac{10^{-6}}{2\pi} \times 1.0 = T_m = 27.87 \text{ Nm}$

With a constant power P_m = 2 kW minimum speed is at maximum torque (set by relief valve), therefore

$$T(max.) \; \omega(min.) = 2000$$

and $n(min.) = \frac{2000}{27.87} \times \frac{1}{2\pi} = 11.43 \text{ rev/s}$

At maximum motor speed

$$2000 = 2\pi \times 20 \times T_m$$

thus $T_m = 15.93 \text{ Nm}$

20% of maximum power $= \frac{3500}{5} = 700 \text{ W}$

Max. torque = 27.87 Nm

and this is maintained constant below a speed of 11.43 rev/s, therefore

$$700 = 2\pi n \times 27.87$$

$$n = 4 \text{ rev/s}$$

52

2. Fig. 2.8 shows a simple hydraulic transmission system having the following details.

Variable Capacity Pump
     Maximum capacity = 80 ml/rev
     Volumetric efficiency = 90%

Electric motor
     Constant running speed = 1000 rev/min
     Overall efficiency = 89%

Variable-capacity hydraulic motor
     Volumetric efficiency = 93%
     Mechanical efficiency = 87%

Pipework
This gives a friction loss equivalent to 8 bar on the high pressure side of the units.

   The overall efficiency of the complete transmission system(excluding the electric motor) is 60%.

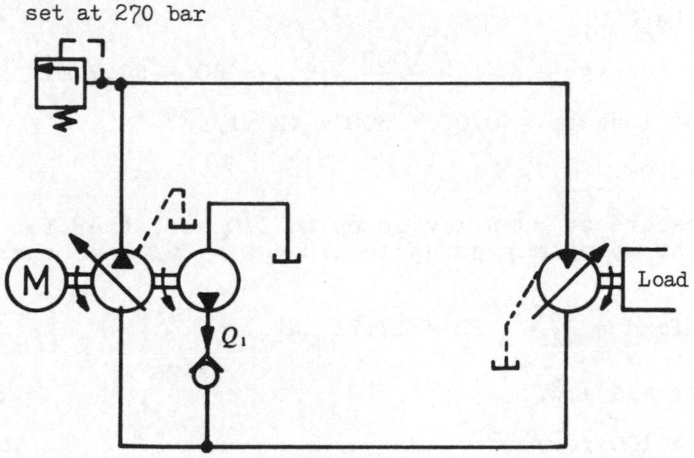

Fig.2.8

Test Condition
   (i)   Pump set at 50% maximum capacity
   (ii) Constant power load of 12 kW on the hydraulic motor
   (iii) All efficiencies to be assumed constant through-out the test
Determine
   (a) the pressure at the pump delivery port
   (b) the leakage flow make-up (Q_1) 1/s

   (c) the permissible percentage increase in load before the relief valve operates

(d) the maximum motor capacity if the minimum running speed of the hydraulic motor, under the given test conditions, is 200 rev/min

(e) the power supplied to the electric motor, and

(f) the mechanical efficiency of the pump under these test conditions.

(a) Pump delivery pressure

$$Q = 0.5 \times \frac{1000}{60} \times 80 \times 0.9 = 600 \text{ ml/s}$$

Pressure at motor inlet $= \dfrac{12 \times 10^3}{Q\eta_{mm}\eta_{vm}}$

$$= \frac{12 \times 10^3}{600 \times 10^{-6} \times 0.93 \times 0.87}$$

$$= 24.7 \times 10^6 \text{ N/m}^2 = 247 \text{ bar}$$

Thus pressure at pump outlet $= 247 + 8 = 255$ bar

(b) Leakage Q_1

Pump leakage $= 0.1 \times \dfrac{1000}{60} \times 0.5 \times 80 = 66.7$ ml/s

Motor leakage $= 0.07 \times 600 = 42$ ml/s

thus $Q_1 = 108.7$ ml/s

(c) Pressure at pump may go up to 270 bar, therefore pressure at motor may go up to $270 - 8 = 262$ bar, therefore

$$\text{New load} = \frac{262}{247} \times 12 = 12.73 \text{ kW}$$

and % increase $= 6.1\%$

(d) $N_m = 200$ rev/min

$$N_m C_m = Q\eta_{vm}$$

therefore

$$\text{Max. } C_m = \frac{600 \times 0.93 \times 60}{200} = 167 \text{ ml/rev}$$

(e) Power input to pump $= \dfrac{12}{0.6} = 20$ kW, therefore

power to electric motor $= \dfrac{20}{0.89} = 22.5$ kW

(f) $\eta_{mp} = \dfrac{\Delta p C_p}{T_p} = \dfrac{\Delta p C_p \omega_p}{T_p \omega_p}$

$$= 255 \times 10^5 \times 0.5 \times \frac{80 \times 10^{-6}}{2\pi} \times 2\pi \frac{1000}{60} \times \frac{1}{20 \times 10^3}$$

$$= 85\%$$

3. A variable-capacity hydraulic pump is used to power a fixed-capacity motor. The details of the two units are as follows.

Pump
       Maximum capacity = 164 ml/rev
       Constant shaft speed = 25 rev/s
The drive is from an electric motor directly coupled to the pump shaft.
       Leakage coefficient = 0.9 ml/bar s
       Mechanical efficiency = 85%

Motor
       Capacity = 65 ml/rev
       Leakage coefficient = 0.9 ml/bar s
       Mechanical efficiency = 85%
The motor load is pure inertia of value 1.0 kg m².

   Neglecting oil compressibility and pipe friction losses determine
   (a) the acceleration of the hydraulic motor when its speed is 33 rev/s and the pump capacity is at 60% maximum
   (b) the power output of the electric motor at the conditions stated in (a)
   (c) the time taken for the load to reach 63% of its maximum speed when the pump capacity is subjected to a step-change from zero to 50% of its maximum value.

   (a) $Q = N_p C_p - \lambda_p \Delta p = N_m C_m + \lambda_m \Delta p$ since C = ml/rev and

N = rev/s. Thus

$$\Delta p = \frac{N_p C_p - N_m C_m}{(\lambda_p + \lambda_m)}$$

$$= \frac{25 \times 0.6 \times 164 \times 10^{-6} - 33 \times 65 \times 10^{-6}}{(0.9 + 0.9)10^{-6}} \quad \text{bar}$$

$$= 178 \text{ bar}$$

$$T_m = J \frac{d\omega_m}{dt}$$

and $\eta_{mm} = \dfrac{2\pi T_m}{\Delta p C_m}$

since C_m = ml/rev, therefore

$$\frac{d\omega_m}{dt} = \frac{\eta_{mm}\Delta p C_m}{2\pi J} = \frac{0.85 \times 178 \times 10^5 \times 65 \times 10^{-6}}{2\pi \times 1.0}$$

$$= 156 \text{ rad/s}^2$$

(b)
$$P(\text{elect.motor}) = \frac{\Delta p Q}{\eta_{op}} = \frac{\Delta p}{\eta_{op}} (N_p C_p \eta_{vp}) = \frac{\Delta p N_p C_p}{\eta_{mp}}$$

$$= \frac{178 \times 10^5 \times 25 \times 0.6 \times 164 \times 10^{-6}}{0.85}$$

$$= 51.4 \text{ kW}$$

(c) From eqn 2.9

$$\left(\frac{J(\lambda_p + \lambda_m)}{\eta_{mm} C_m^2} D + 1 \right) \omega_m = \omega_p \frac{C_p}{C_m}$$

where $D = d/dt$. (Note that C_m and C_p are capacities per radian.)

Solution for step input to

$$(\tau D + 1) \, \omega_m = \omega_p \frac{C_p}{C_m}$$

is $\omega_m = \omega_p \dfrac{C_p}{C_m} \left(1 - e^{-t/\tau} \right)$

when $t > 5\tau$

$$\omega_m \simeq \omega_p \frac{C_p}{C_m} = \text{max. speed}$$

for $\omega_m = 63\%$ maximum

$$e^{-t/\tau} = 37\% \text{ i.e. } 0.37 \, e^{t/\tau} = 1$$

thus $e^{t/\tau} = 2.7$

i.e. $t \simeq \tau$

$$= \frac{1 \, (0.9 + 0.9) 10^{-6}}{10^5 \times 0.85 \left[\frac{65}{2\pi} \times 10^{-6} \right]^2} \text{ s}$$

$$= 19.8 \times 10^{-2} \text{ s}$$

Note: This time to reach 63% of maximum speed will be the

same for any size of step input. It is only the actual
value of the speed that is affected by the pump capacity
employed.

4.   A proposed transmission system is to consist of a
variable-capacity pump supplying oil to a fixed-capacity
motor. The motor is to be directly coupled to a pure
inertia load of 2 kg m^2. The total volume of oil on the
high pressure side of the system is calculated to be 2
litres and the leakage coefficient is estimated at 0.4 ml/
bar s for the pump and the motor separately. Pressure
losses between the pump and motor are to be neglected and
a value of 17k bar is to be used for the effective bulk
modulus of the fluid. The pump is to run at 12 rad/s and
have a maximum capacity of 18 ml/rad. The motor is to
have a capacity of 12 ml/rad and an estimated mechanical
efficiency of 90%.

   Determine
   (a) the damped natural frequency of the system in Hz
   (b) the damping ratio that exists
   (c) the value of the load inertia that will just give
critical damping assuming all other parameters remain unchanged

   Using eqn 2.8 obtained by standard approach for (a)
and (b)

$$\left[D^2 + \frac{B'(\lambda_p + \lambda_m)}{V_0} D + \frac{\eta_m B' C_m^2}{JV_0} \right] \omega_m = \frac{\eta_m B' C_m^2}{JV_0} \omega_p \frac{C_p}{C_m}$$

gives

$$(D^2 + 6.8D + 58) \, \omega_m = 58 \times 10^6 \, C_p \qquad (2.21)$$

therefore $\omega_n^2 = 58$ giving $\omega_n = 7.6$ rad/s undamped natural
frequency and $2\zeta\omega_n = 6.8$ giving $\zeta = 0.448$ (damping ratio).
Therefore $\omega_d = \omega_n \sqrt{(1 - \zeta^2)} = 6.8$ rad/s (natural damped
frequency)

i.e. $n_d = 1.08$ Hz

giving the periodic time of damped oscillation as 0.924 s

   (c) For critical damping $\zeta = 1.0$ i.e. minimum damping
with no oscillations. Therefore

$$\frac{2\lambda}{2C_m} \sqrt{\left(\frac{JB'}{\eta_m V_0} \right)} = 1.0$$

substituting the appropriate values gives $J = 0.95$ kg m^2.
i.e. reducing the inertia from 2 kg m^2 increases the damping ratio from 0.448 to 1.0. This only applies to a viscous friction free system.

57

5. A transmission system is represented by eqn 2.8. If the system is subjected to a step input from zero to $\omega_p C_p/C_m = \omega_0$, determine the relationship between the damping ratio ζ and the ratio of successive amplitudes ω_0/ω_1, ω_1/ω_2, etc., as indicated in fig. 2.9.

Eqn 2.8 may be written as

$$(D^2 + 2\zeta\omega_n D + \omega_n^2)\,\omega_m = \omega_n^2\,\omega_p\,\frac{C_p}{C_m}$$

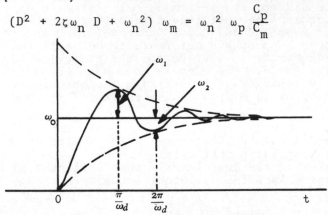

Fig.2.9

The solution for a step input ω_0 for $t \geq 0$ is

$$\omega_m = \omega_0\left[1 - e^{-\zeta\omega_n t}\left(\cos\omega_d t + \frac{\zeta\omega_n}{\omega_d}\sin\omega_d t\right)\right]$$

The decrement

$$\left|\frac{\omega_0}{\omega_1}\right| = \left|\frac{\omega_1}{\omega_2}\right| = \frac{e^{-\zeta\omega_n t}}{e^{-\zeta\omega_n\left(t + \frac{\pi}{\omega_d}\right)}} = e^{\pi\zeta\frac{\omega_n}{\omega_d}} = e^{\pi\zeta/\sqrt{(1 - \zeta^2)}} \qquad (2.22)$$

therefore

$$\log_e (\text{decrement}) = \pi\zeta/\sqrt{(1 - \zeta^2)} = \lambda \text{ (say)}$$

and $\zeta = \dfrac{\lambda}{\sqrt{(\pi^2 + \lambda^2)}}$ \qquad (2.23)

Hence by experimentation to determine the ratio of any two successive amplitudes, and hence λ,,the damping ratio of the system may be computed directly from eqn 2.23. Note that when $\lambda = \pi$ then $\zeta = 0.7$.

Fig. 2.10a shows the variation of ζ with λ (i.e. $\log_e|\omega_0/\omega_1|$) or any two other successive amplitudes. Fig.

58

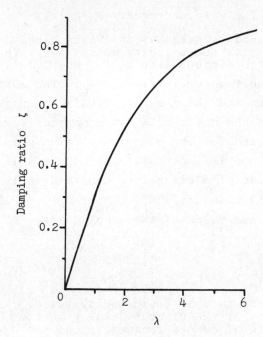

Fig.2.10a

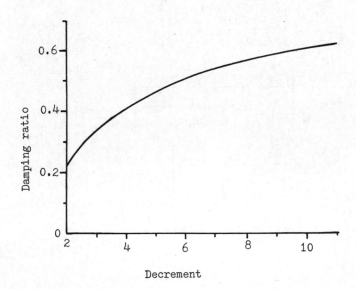

Fig.2.10b

59

2.10b shows variations of ζ with decrement $|\omega_0/\omega_1|$.

6. A hydraulic transmission system has an inertia load J coupled directly to the constant capacity motor shaft. This shaft is also subjected to viscous damping $f\omega_m$ where f is the damping torque per unit angular velocity ω_m. The motor capacity is C_m per radian and the mechanical efficiency is η_m. The leakage rate of the motor is λ_m (volumetric flow rate per unit pressure) and for the pump is λ_p. The pump speed is constant at ω_p and its instantaneous capacity is C_p per radian. The volume of fluid before compression is V_0 and the effective bulk modulus is B'.

(a) Show that the minimum value of the damping ratio ζ occurs when

$$\frac{JB'}{V_0} = \frac{f}{\lambda_p + \lambda_m}$$

and determine the value of this minimum damping ratio.

(b) Discuss the effects of the viscous damping on the natural undamped and natural damped frequencies of the system.

The standard approach will yield eqn 2.16 as the governing equation of the system. From this it can be seen that

$$\omega_n = C_m \sqrt{\left[\frac{B'\eta_m}{JV_0}\left(1 + \frac{f(\lambda_p + \lambda_m)}{\eta_m C_m^2}\right)\right]} \qquad (2.18)$$

$$\text{and } \zeta = \frac{(\lambda_p + \lambda_m)}{2C_m}\sqrt{\left[\frac{JB'}{\eta_m V_0}\frac{1}{1 + \dfrac{f(\lambda_p + \lambda_m)}{\eta_m C_m^2}}\right]}$$

$$+ \frac{f}{2C_m}\sqrt{\left[\frac{V_0}{\eta_m JB'}\frac{1}{1 + \dfrac{f(\lambda_p + \lambda_m)}{\eta_m C_m^2}}\right]} \qquad (2.19)$$

(a) For minimum ζ the differential

$$\frac{d\zeta}{d\left(\dfrac{JB'}{V_0}\right)} = 0$$

This reveals

$$\frac{JB'}{V_0} = \frac{f}{\lambda_p + \lambda_m} \tag{2.24}$$

i.e. $\tau\omega_n^2 = \frac{f}{J}$

Substituting eqn 2.24 into eqn 2.19 gives

$$\zeta(min.) = \frac{1}{\sqrt{\left[1 + \dfrac{\eta_m C_m^2}{f(\lambda_p + \lambda_m)}\right]}} \tag{2.25}$$

and ω_n at this value of ζ is given by

$$\omega_n \ (at \ \zeta \ (min.) = \frac{f}{J}\sqrt{\left[1 + \frac{\eta_m C_m^2}{f(\lambda_p + \lambda_m)}\right]} \tag{2.26}$$

i.e. $\omega_n = \frac{f}{J}\sqrt{\left(1 + \frac{J}{f\tau}\right)}$

Eqn 2.25 × eqn 2.26 reveals

$$\zeta(min.) \ \omega_n' = \frac{f}{J}$$

where $\omega_n' = \omega_n$ at $\zeta(min.)$.

(b) Rewriting eqn 2.18 as

$$\omega_n = C_m\sqrt{\left\{\frac{B'\eta_m}{JV_0}\left[1 + \frac{f(\lambda_p + \lambda_m)}{\eta_m C_m^2}\right]\right\}} \tag{2.18}$$

Now the natural frequency without viscous damping is given by eqn 2.12

$$\omega_n = C_m\sqrt{\left(\frac{B'\eta_m}{JV_0}\right)} \tag{2.12}$$

Comparing eqn 2.12 and 2.18 shows that the presence of viscous damping has increased the natural frequency of the system. Expressing eqn 2.18 as

$$\omega_n = C_m\sqrt{\left(\frac{B'\eta_m}{JV_0}\right)}\left[1 + \frac{f(\lambda_p + \lambda_m)}{\eta_m C_m^2}\right]^{\frac{1}{2}}$$

clearly shows that the increase is

$$\left[1 + \frac{f(\lambda_p + \lambda_m)}{\eta_m C_m{}^2}\right]^{\frac{1}{2}} - 1$$

i.e. a 10% increase requires that

$$f = 0.21 \frac{\eta_m C_m{}^2}{(\lambda_p + \lambda_m)}$$

Damped natural frequency $\omega_d = \omega_n \sqrt{(1 - \zeta^2)}$, therefore

$$\omega_d{}' = \sqrt{\left\{\frac{C_m{}^2 B' \eta_m}{JV_0} - \left[\frac{(\lambda_p + \lambda_m)B'}{2V_0} - \frac{f}{2J}\right]^2\right\}} \qquad (2.27)$$

compared with that for no viscous friction f, of

$$\omega_d = \sqrt{\left\{\frac{C_m{}^2 B' \eta_m}{JV_0} - \left[\frac{(\lambda_p + \lambda_m)B'}{2V_0}\right]^2\right\}} \qquad (2.28)$$

i.e. $\omega_d = \omega_n \sqrt{\left[1 - \left(\frac{\omega_n \tau}{2}\right)^2\right]}$

An increase of ω_d resulting from the introduction of viscous damping means that ω_d from eqn 2.27 must be greater
than ω_d from eqn 2.28, therefore

$$\left[\frac{(\lambda_p + \lambda_m)B'}{2V_0} - \frac{f}{2J}\right]^2 < \left[\frac{(\lambda_p + \lambda_m)B'}{2V_0}\right]^2$$

i.e. $\dfrac{f}{4J^2} < \dfrac{(\lambda_p + \lambda_m)B'f}{2JV_0}$

i.e. $\dfrac{f}{(\lambda_p + \lambda_m)} < \dfrac{2JB'}{V_0}$ \qquad (2.29)

hence

$$\frac{JB'}{V_0} < \frac{f}{(\lambda_p + \lambda_m)} < \frac{2JB'}{V_0}$$

from statements 2.24 and 2.29

It should be noted that substituting the conditions for
minimum ζ from eqn 2.24 into eqn 2.27 gives eqn 2.28.

7. The variable delivery pump in fig. 2.11 is pressure compensated so that the delivery is reduced as the pressure rises above a preset value.

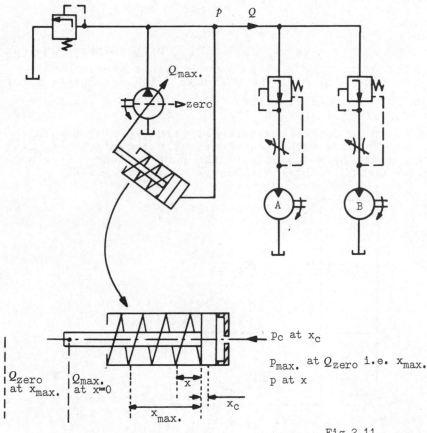

Fig.2.11

Describe the action of this pump arrangement and show that

$$P_{max} = \frac{Q_{max}(p_{max})^2}{4(p_{max} - p_c)}$$

where P_{max} = maximum pump delivery power

Q_{max} = maximum pump delivery flow

p_{max} = maximum pump delivery pressure

p_c = cut-off pressure i.e. the pressure at which the pump delivery begins to decrease as the pressure increases (set by spring pre-compression x_c).

63

and $p_c \leq \frac{1}{2} p_{max}$

If $p_c > \frac{1}{2} p_{max}$ determine the maximum power and the pressure at which this occurs.

Action

The pump stroke lever is set to Q_{max} by the spring-loaded cylinder unit. The spring is such that a pressure rise from 0 to p_c will move the piston x with negligible alteration in pump delivery flow. If the motors A and B are now adjusted for speed so that Q_{max} is no longer required then the pump outlet pressure will tend to rise towards the relief valve value. However, as p increases the spring moves by x and reduces the pump delivery to balance the new demand. This is illustrated in fig. 2.12.

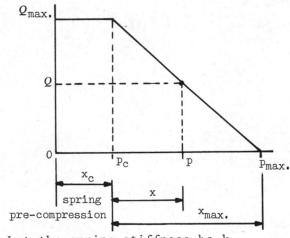

Fig.2.12

Let the spring stiffness be k

At $Q = Q_{max}$ $\quad p_c = kx_c$
At $Q = Q$ $\quad p = k(x_c + x)$
At $Q = 0$ $\quad p_{max} = k(x_c + x_{max})$

therefore

$$Q = Q_{max} - \frac{x}{x_{max}} Q_{max} = Q_{max}\left(1 - \frac{p - p_c}{p_{max} - p_c}\right) \quad (2.30)$$

Now, power

$$P = pQ$$
$$= pQ_{max}\left(\frac{p_{max} - p}{p_{max} - p_c}\right) \quad (2.31)$$

64

for max. power $dP/dp = 0$, therefore

$$p = \frac{1}{2} P_{max}$$

and $P_{max} = \dfrac{Q_{max}(p_{max})^2}{4(p_{max} - p_c)}$ \hfill (2.32)

If $p_c > \frac{1}{2} p_{max}$ then the power will be pQ_{max} up to $p_c Q_{max}$ as illustrated in fig. 2.14. Beyond a pressure p_c the power will decrease as fig. 2.14 also shows.

Construction of fig. 2.13

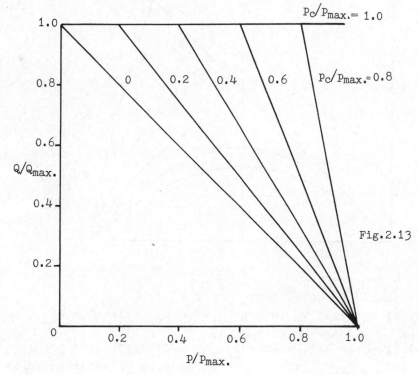

Fig.2.13

The flow rate Q may be expressed as (from eqn 2.30)

$$Q = Q_{max} \left(\frac{p_{max} - p}{p_{max} - p_c} \right)$$

hence

$$\frac{Q}{Q_{max}} = \frac{1 - p/p_{max}}{1 - p_c/p_{max}} \hfill (2.33)$$

For various values of p_c/p_{max} the dimensionless flow

rate (eqn 2.33) is plotted as in fig. 2.13. From eqns 2.31 and 2.33

$$\frac{P}{P_{max}Q_{max}} = \frac{p}{P_{max}}\left[\frac{Q}{Q_{max}}\right] \qquad (2.34)$$

The dimensionless power (eqn 2.34) is plotted in fig. 2.14.

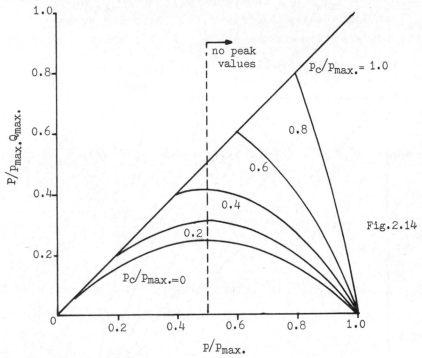

Fig.2.14

It should be noted that

(a) in eqn 2.33 while $p/p_{max} \le p_c/p_{max}$ then $Q = Q_{max}$ due to the nature of the system (see fig. 2.13).

(b) from eqn 2.34 when $p/p_{max} > 0.5$ the maximum power output is $p_c Q_{max}$. i.e. there is no peak value as given by eqn 2.32.

FURTHER EXAMPLES

1. A power transmission circuit consists of a variable-capacity positive displacement pump, fixed-capacity positive displacement hydraulic motor and associated pipes and fittings. The pump is driven by an electric motor which runs at a constant speed of 1000 rev/min. The following details apply to the circuit components

```
        Max pump capacity = 164 ml/rev
        Pump volumetric efficiency = 92%
        Hydraulic motor volumetric efficiency = 90%
        Hydraulic motor mechanical efficiency = 78%
```

Hydraulic losses in pipes and fittings and valves between the pump and motor are equivalent to 10 bar. The hydraulic motor is subjected to a constant torque load equivalent to 3.75 kW at a maximum speed of 2000 rev/min.

Determine for maximum load conditions the capacity of the hydraulic motor being used and the pump delivery pressure assuming suction pressure is atmospheric.

Select one of the following electric motors to power the unit, making any reasonable assumptions and stating what these might be.

```
        Electric motor         A    B    C    D    E    F
        Rating kW (nett output)  65   70   75   80   85   90
```

(68 ml/rev, 220 bar, motor D.)

2. A hydrostatic transmission system consists of a variable-capacity pump and a variable-capacity motor. The loss of pressure between the pump and the motor in the high pressure line is given by the empirical formula $\Delta p = 0.12 \times 10^{-5} Q^2$ where Δp_f is expressed as bar and Q is the oil flow rate in ml/s.

Pump Data

```
        Shaft speed = 1500 rev/min
        Maximum capacity = 100 ml/rev
        Volumetric efficiency = 97%
        Overall efficiency = 70%
```

Motor Data

```
        Volumetric efficiency = 96%
        Mechanical efficiency = 83%
```

Initially the pump capacity is set to zero and the motor capacity is set to some fixed value. A constant torque load of 21.5 Nm is applied to the motor shaft and the speed is increased by altering the pump capacity only. The maximum motor speed obtainable is 4000 rev/min.

Determine the motor capacity being employed and the pressure drop across the motor at 2000 and 4000 rev/min motor speed.

The motor speed is now reduced from its maximum value of 4000 rev/min by altering the motor capacity only, while the load is maintained constant at 7.1 kW. Determine the values of the motor capacity and the pump delivery pressure at a motor speed at 2500 rev/min.

If, finally, the motor capacity is fixed at this new value and the pump capacity is altered to reduce the speed of the motor to 1000 rev/min while the same constant power load is maintained, determine the new pump delivery pressure.

(34.8 ml/rev, 47 bar, 55.7 ml/rev, 37 bar, 93.4 bar)

3. In a flow controlled hydraulic system the maximum circuit pressure is 83 bar. The maximum capacity of the variable capacity pump is 115 ml/rev and the capacity of the fixed capacity motor is 148 ml/rev. The pump is directly coupled to an electric motor and is driven at a constant speed of 1000 rev/min. If the overall efficiency of the pump and hydraulic motor is 84 per cent each and mechanical efficiency 90 per cent each, determine, neglecting losses in piping valves, etc.

(a) the maximum motor speed and the power developed at this speed.
(b) the torque supplied to the pump from the electric motor under these conditions.
(c) the leakage coefficients for the pump and hydraulic motor.
(d) the percentage loss of power in the whole system.

((a) 11.2 rev/s 12.45 kW (b) 168.5 Nm (c) λ_p = 0.157 ml/
bar s λ_m = 0.150 ml/bar s (d) 29.4%)

4. An experimental flow controlled hydraulic circuit consists of a variable delivery pump of maximum capacity 90 ml/rev driven at a constant speed of 960 rev/min and a maximum delivery of 1440 ml/s of oil, supplying a constant displacement motor of capacity 49 ml/rev. Under maximum conditions the power absorbed by the pump is 3.75 kW and the maximum circuit pressure is 24 bar. Determine

(a) volumetric, mechanical and overall efficiencies of the pump and its leakage coefficient.
(b) maximum speed of motor assuming it has the same leakage coefficient as the pump but not the same volumetric efficiency.
(c) volumetric and mechanical efficiencies for the motor given, overall efficiency of 76%.
(d) power delivered by the motor at maximum speed
(e) overall efficiency of plant.

((a) 99.7%, 93%, 92.5% (b) 1747 rev/min (c) 99.6%, 76.4%
(d) 2.65 kW (e) 71.3%)

5. In a flow controlled hydraulic system (variable-stroke pump driving fixed-stroke motor) a maximum output torque

68

of 13.6 Nm and maximum output speed of 1000 rev/min are
required. The maximum working pressure in the system is
to be 70 bar and the pump is driven by an electric motor
at a constant speed of 1500 rev/min. Determine the minimum suitable value for motor capacity, maximum pump capacity and power rating of the electric motor, assuming
that the pump and hydraulic motor each have a mechanical
efficiency of 95% and the leakage coefficient of the pump
is 0.24 ml/bar s an d of the motor is 0.40 ml/bar s.

(13 ml/rev, 10.4 ml/rev, 1.9 kW)

6. In a flow controlled hydraulic circuit the motor capacity is 33 ml/rev. The pump output (max) is 610 ml/s and
its constant speed is 960 rev/min with 3.75 kW being supplied from an electric motor. If the overall efficiency
is 84% and mechanical efficiency is 90% for both pump and
motor determine

(a) the maximum circuit pressure
(b) the maximum motor speed and torque at this speed
(c) pump maximum capacity
(d) coefficients of leakage of pump and motor

((a) 51.6 bar (b) 24.3 Nm 1035 rev/min (c) 40.5 ml/rev (d)
0.8 ml/bar s)

7. A variable-capacity pump and a variable-capacity motor
are linked hydraulically by pipework and valves which may
be considered to give a pressure loss of 10% of the pump
delivery pressure. The motor is subjected to a constant
torque load of 28 Nm. The pump and motor both have maximum capacities of 50 ml/rev and the pump is driven at a
constant speed of 1000 rev/min.

When in operation, speed control is effected by first
fixing the motor at maximum capacity and gradually increasing the pump capacity from zero to maximum value.
The speed is then further increased by reducing the motor
capacity while the pump remains at its maximum setting.
Assuming 100% efficiency for both pump and motor, determine

(a) the capacity of the motor for a speed of 2500 rev/
min.
(b) the relief valve setting (if this is situated near
the pump) to limit the motor speed to 5000 rev/min under
this constant torque load.

If the pump is set at 50% of its maximum capacity and
speed variation is obtained by adjusting the motor capacity only, what new relief valve setting would be necessary to limit the motor speed to 5000 rev/min with a constant torque load of 11.5 Nm?

((a) C_m = 20 ml/rev, (b) 200 bar, 160 bar)

69

8. A pump and motor unit consists of a variable-capacity pump and a fixed-capacity motor with associated pipes and valves. The motor details are

    Capacity/rev = 50 ml
    Leakage coefficient = leakage coefficient of pump
    Mechanical efficiency = 80%
    Overall efficiency = overall efficiency of pump

Pump details are

    Maximum capacity = 84 ml/rev
    Constant shaft speed = 1500 rev/min
    Leakage rate = 22 ml/s under test conditions stated

Test conditions

Hydraulic motor develops 3.75 kW with a constant torque load of 15 Nm. Pressure drop in circuit between pump and motor = 20% of pump delivery pressure.

    Calculate
    (a) Flow rate between pump and motor.
    (b) Volumetric efficiencies of pump and motor assuming leakage volume/min = $\lambda \Delta p$, and the leakage coefficient = λ.
    (c) Percentage of maximum pump capacity being employed.
    (d) Power and torque inputs to pump.

        ((a) 17.5 1/s (b) 99% (c) 86% (d) 7.46 kW, 47.5 Nm)

9. A circuit consists of a fixed-capacity pump, variable-capacity motor and associated pipes and fittings. The pressure drop between the pump and the motor on the high pressure side is 17 bar. The pump and motor details are

Pump   capacity = 82 ml/rev,
          speed = 1500 rev/min
          volumetric efficiency = 90%
          mechanical efficiency = 84%

Motor   maximum capacity = 66 ml/rev,
          volumetric and mechanical efficiencies as for pump

The relief valve, adjacent to the pump, is set at 135 bar.

    If the motor is subjected to a constant torque load of 34 Nm, determine

    (a) the minimum motor speed and the pressure drop across the motor at this speed
    (b) the maximum motor speed and the associated motor capacity
    (c) the theoretical maximum power that can be transmitted and the speed range associated with it.

((a) 1520 rev/min 39 bar (b) 4740 rev/min 21 ml/rev (c) 16.8 kW, 1520/4740 rev/min)

10. A transmission system is to consist of a variable-capacity pump and a fixed-capacity motor. The circuit is to deliver a constant power of 45 kW over a speed range of 4000 to 8000 rev/min. The maximum permissible pressure is 172 bar and the efficiency of transmission of the pipes and valves between the pump and motor is estimated as 60%. Assuming a volumetric efficiency of 90% and an overall efficiency of 80% for both pump and motor, determine suitable maximum capacities for both units, maximum oil flow rate, minimum pressure at the pump delivery and the motor torque associated with this pressure. Assume a pump shaft speed of 1460 rev/min.

Since the speed range quoted above is high for hydraulic motors, determine the new capacity requirements of units employed in the circuit shown in fig. 2.15 to perform exactly the same duty.

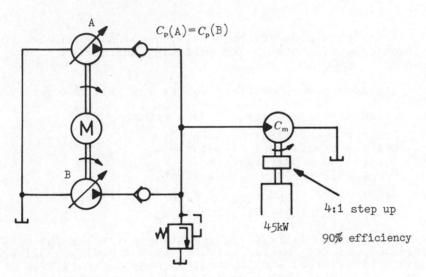

$C_p(A) = C_p(B)$

4:1 step up

45kW

90% efficiency

Fig.2.15

(C_m = 73 ml/rev, C_p = 490 ml/rev, 10.8 l/s, 86.5 bar, 31.6 bar; C_m = 18.3 ml/rev, C_p = 61.5 ml/rev)

11. A fixed-capacity hydraulic pump delivers oil via a variable restrictor to a fixed-capacity hydraulic motor. The pump capacity is 115 ml/rev, speed 1000 rev/min and volumetric efficiency 95%. The motor capacity is 164 ml/rev, volumetric efficiency 95% and mechanical efficiency 80%. The relief valve setting is 69 bar and the motor load is a constant torque of 56.5 Nm. The flow through the restrictor is given by the expression

$$Q \left[\frac{ml}{s}\right] = 1.29 \ a_0 \sqrt{(\Delta p_0)}$$

71

where a_0 = restrictor orifice area (mm²); a_0(max) = 322 mm², Δp_0 = restrictor orifice pressure drop (bar)

Determine the maximum motor speed and the motor speed when a_0 = 129 mm².

(634 rev/min, 398 rev/min)

12. A hydraulic pump and a motor have identical swept volumes (C) and leakage coefficients (λ) and run at the same speed (N). If the pressure change across each is identical at Δp show that the volumetric efficiencies of the two units are within 1% of each other provided $\lambda \Delta p < 0.1$ NC.

At what value of pump volumetric efficiency will the motor volumetric efficiency be 80%?

(83.2%)

13. A variable-capacity pump is hydraulically coupled to a fixed-capacity motor. The following data refer to the system

Variable pump capacity = C_p (at any instant) per radian
Fixed motor capacity = C_m per radian
Leakage coefficient of pump = λ
Leakage coefficient of motor = λ
Motor load = pure inertia J
Pump shaft speed = ω_p (constant)
Motor shaft speed = ω_m at pump capacity C_p
Motor mechanical efficiency = η_m

Establish the differential equation governing the motor speed response to any variation of pump capacity. Solve the equation for a 'step' change of pump capacity from zero to Z. After what time interval is the motor speed equal to

$$\left(\frac{e-1}{e}\right)\frac{\omega_p Z}{C_m} \quad ?$$

$$(t = 2J/\lambda \eta_m C_m^2)$$

14. Fig. 2.16a shows a hydraulic transmission system to which the following data apply

Friction pressure loss between the pump and motor $\Delta p_f = 3 \times 10^{-3}$ Q where Δp_f = bar and Q = ml/s

p_1 = 55 bar gauge
p_2 = 50 bar gauge

72

$C_m = 75$ ml/rev

motor volumetric efficiency = 93%
motor mechanical efficiency = 78%
C_p(max) = 125 ml/rev
N_p(constant) = 16.3 rev/s
pump volumetric efficiency = 90%

Determine

(a) the motor speed.
(b) the power being delivered by the motor.
(c) the percentage of the maximum pump capacity being used.

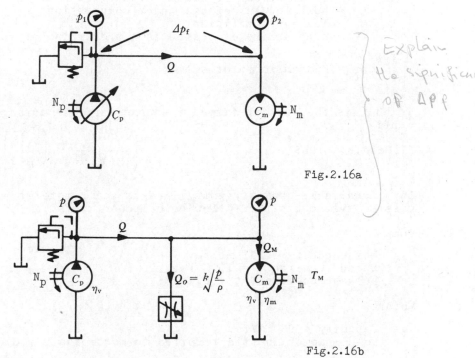

Fig.2.16a

Explain
the significant
of Δp_f

Fig.2.16b

For the system shown in fig. 2.16b establish the relationship

$$N_m = AC_p - B\sqrt{T_m}$$

where A and B are constants. Note p is less than the relief valve setting.

((a) 20.6 rev/s (b) 6 kW (c) 91%)

15. A hydraulic transmission consists of a constant speed electric motor driving a variable-displacement pump. The

pump supplies oil to a fixed-displacement motor. Prove that the relationship between pump stroke and hydraulic motor speed is given by

$$\frac{n_m}{\theta_p} = n_p V_p \left[\frac{2\pi}{V_m} \frac{VI}{N} D^2 + \frac{2\pi}{V_m} ILD + \frac{V_m}{2\pi} \right]^{-1}$$

where n_p = pump speed

V_p = pump displacement per rev at maximum stroke

V_m = motor displacement per rev

V = volume of oil under compression

I = total effective moment of inertia

N = bulk modulus of oil under compression

L = leakage coefficient

θ_p = pump stroke expressed as a proportion of full stroke

n_m = hydraulic motor speed

D = d/dt

What is the natural damped frequency of this arrangement?

$$\left(\frac{V_m N}{2\pi V} \sqrt{\left[\frac{V}{NI} - \left(\frac{\pi L}{V_m} \right)^2 \right]} \right)$$

16. A variable-capacity pump/fixed-capacity motor transmission system has the following component details

Pump

    Max capacity = 131 ml/rev
    Speed (constant) = 1000 rev/min
    Leakage coefficient = 0.18 ml/bar s

Motor

    Capacity = 98 ml/rev
    Load + motor inertia referred to motor shaft = 0.62 kg m^2
    Mechanical efficiency = 87%
    Leakage coefficient = 0.20 ml/bar s

    Determine

(a) The time taken for the motor speed to reach 95% of its steady state value if the pump is subjected to a 'step' capacity change from zero to 50% of maximum. What are the values of the speed and acceleration at this instant?

(b) The steady state speed lag when the pump capacity is increased at a uniform rate from zero to maximum in three seconds.

((a) 3.4 × 10^{-2} s, 660 rev/min, 0.516 rad/s)

17. A fixed-capacity pump supplies oil to a variable capacity motor which delivers a constant power output of 15 kW. The following data apply to the system

    Pump capacity = 131 ml/rev
    Pump speed = 1000 rev/min
    Pump volumetric efficiency = 90%
    Motor maximum capacity = 82 ml/rev
    Motor volumetric efficiency = 95%
    Motor overall efficiency = 80%
    Loss of pressure between pump and motor = 0.1p, where p is the pump delivery pressure
    Overall efficiency of the complete transmission system is 50%

   Calculate

  (a) the pump delivery pressure
  (b) the mechanical efficiency of the pump
  (c) the maximum output torque of the motor and the speed at which this torque is delivered

      ((a) 105 bar (b) 77% (c) 104 Nm at 1370 rev/min)

18. A series of tests on a pure inertia loaded hydraulic transmission rig produced the following results

  (a) A plot of the output speed on a time base indicated a second order differential equation governing the system.
  (b) The ratio of the first overshoot to the step input was 4.33×10^{-2}.
  (c) The damped natural frequency was 2.0 Hz.
  (d) The viscous friction was negligible.
  (e) The effective bulk modulus of the fluid was 16×10^3 bar.
  (f) The combined leakage coefficients for the system was 0.7 ml/bar s.
  (g) The motor capacity was 50 ml/rev and its mechanical efficiency was 88%.

    Determine the percentage change of inertia load at motor shaft necessary to produce critical damping and the actual damping torque per unit angular velocity.

    What was the torsional stiffness of the system?

(% increase of J = 104%, damping torque = 49.5 Nm s/rad, torsional stiffness = 610 Nm/rad)

19. The following data refer to a transmission system where the pipe friction losses may be neglected

Pump	Motor
C_p(max) = 20 ml/rad	C_m(fixed) = 20 ml/rad
N_p(constant) = 16 rev/s	λ_m = 0.4 ml/bar s

λ_p = 0.3 ml/bar s $\qquad\qquad$ η_{mm} = 90%

Volume of oil on high pressure side = 2.5 litres
Effective bulk modulus = 16.5×10^3 bar
Load (pure inertia) = 3 kg m^2

If the pump capacity is subjected to a step change
from zero to 60% of its maximum value show that the instantaneous motor speed is approximately 30% of the pump
speed after a time lapse equal to twice the periodic time
of damped oscillation of the system.

20. A fixed-capacity motor/variable-capacity pump transmission system is governed by the differential equation

$$(2.04 \times 10^{-2} D^2 + 10^{-1} D + 1) \omega_m = \omega_p \frac{C_p}{C_m}$$

where D = d/dt
ω_m = motor shaft speed (rad/s)
ω_p = pump shaft speed (rad/s)
C_m = motor capacity (m^3/rad)
C_p = pump capacity (m^3/rad)

(a) Determine the damping ratio and damped natural frequency of the system.
(b) If C_m = 0.5 C_p and ω_p = 0 at t = 0 and ω_p = 100t
when t ≥ 0 determine the steady state lag of ω_m relative
to $2\omega_p$.

(0.35 = ζ, 6.55 rad/s = ω_d, lag = 20 rad/s)

21. The inertia load directly coupled to the shaft of a
hydraulic motor has the value of 1.5 kg m^2. The shaft is
subjected to viscous damping. The motor capacity is constant at 10 ml/rad and the mechanical efficiency is 85%.
The pump and motor both have leakage coefficients of 0.35
ml/bar s; the bulk modulus of the oil is 16.5×10^3 bar
and the volume of oil under compression is 1.8 litres.

Determine

(a) the value of the viscous damping if this gives the
natural frequency of the system a 5% increase over that
when there is no viscous damping present and
(b) the natural damped frequency of the system when the
viscous friction is that calculated in part (a).

(1.22 Nm s/rad 6.65 rad/s)

22. A variable-capacity pump/fixed-capacity motor system
is to be used to provide an output shaft rotation θ_m at
the motor resulting from an input at the pump described

by the parameter $\omega_p C_p / C_m$ where the symbols have their normal significance. Show that the transfer function for the system is

$$\left(\frac{\theta_m}{\frac{\omega_p C_p}{C_m}}\right)(s) = \frac{1}{s\left(\frac{1}{\omega_n^2}\, s^2 + \frac{2\zeta}{\omega_n}\, s + 1\right)}$$

where $\omega_n^2 = \eta_m B' C_m^2 / (JV_0)$

$\quad\quad \eta_m$ = mechanical efficiency of the motor

$\quad\quad B'$ = effective bulk modulus for oil volume V_0

23. Three hydraulic motors are used to provide the drive on the three axes of an N.C. machine. The motors are supplied from a single variable-capacity pump which is pressure compensated. The maximum flow from the pump is 100 1/min and this is required while the total load pressure, seen by the pump, rises to 40 bar. Beyond this pressure the flow will reduce in a linear manner to the stall pressure of 200 bar.

   Determine

   (a) The power of a simple fixed-capacity pump that would provide these pressures and flows.
   (b) The maximum power required by the compensating pump detailed and the flow at which this occurs.
   (c) The maximum power at the pump if the delivery is to remain constant at its maximum value until the presssre is 120 bar.
   (d) The pump flow rate when the pressure is 160 bar assuming maximum flow rate is available up to 120 bar.

((a) 33.33 kW (b) 10.4 kW, 1.04 1/s (c) 20 kW (d) 0.83 1/s)

3 VALVE-CONTROLLED SYSTEMS

Valves, within a system, are used to control

(a) pressure level
(b) direction of fluid flow
(c) rate of fluid flow
(d) sequential events based on pressure or position

This chapter is concerned with the control of rate of fluid flow and will deal mainly with valve and ram units (a ram unit being a piston and cylinder).

FLOW THROUGH A SINGLE ORIFICE

Consider a small sharp-edged orifice with its axis horizontal as indicated in fig. 3.1, and an incompressible fluid.

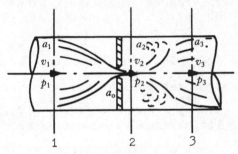

Fig.3.1

p_1 = pressure at point 1 where it is uniform across the diameter
a_1 = area at point 1; v_1 = mean velocity at 1
p_2 = pressure at point 2 which is the vena contracta i.e. point of minimum area where the jet sides are parallel
a_2 = area at point 2; v_2 = mean velocity at 2
Δp_f = the pressure loss between 1 and 2 due to friction, shock, etc.
a_o = orifice area
ρ = fluid density

Applying the steady-flow energy equation between point 1 and point 2 gives

$$\frac{p_1}{\rho} + \frac{v_1{}^2}{2} = \frac{p_2}{\rho} + \frac{v_2{}^2}{2} + \frac{\Delta p_f}{\rho}$$

The continuity equation gives

$$a_1 v_1 = a_2 v_2$$

Hence

$$v_2 = \left[1 - \left(\frac{a_2}{a_1}\right)^2 \right]^{-\frac{1}{2}} \left[\frac{2}{\rho} (p_1 - p_2 - \Delta p_f) \right]^{\frac{1}{2}}$$

and the volumetric flow rate Q is given by

$$Q = a_2 \left[1 - \left(\frac{a_2}{a_1}\right)^2 \right]^{-\frac{1}{2}} \left[\frac{2}{\rho} (p_1 - p_2 - \Delta p_f) \right]^{\frac{1}{2}}$$

If a_2 may be expressed as $a_2 = C_1 a_0$ where C_1 is a factor which is assumed constant, and if $p_1 - p_2 - \Delta p_f$ is written as $(p_1 - p_2)[1 - \Delta p_f/(p_1 - p_2)]$ then

$$Q = \frac{C_1 \left[1 - \frac{\Delta p_f}{p_1 - p_2} \right]^{\frac{1}{2}}}{\left[1 - C_1^2 \left(\frac{a_0}{a_1}\right)^2 \right]^{\frac{1}{2}}} \times a_0 \left[\frac{2}{\rho}(p_1 - p_2) \right]^{\frac{1}{2}}$$

Since C_1 and Δp_f are not easily determined then, provided a_1 is considerably greater than a_0, this expression may be written as

$$Q = C_d a_0 \sqrt{\left[\frac{2(p_1 - p_2)}{\rho} \right]} \qquad\qquad (3.1)$$

C_d is called a coefficient of discharge for the orifice and varies with the Reynolds number of the flow. Variations of the definition of this coefficient exist but for fluid power work this particular definition is found to be most convenient. The value of C_d varies considerably with Reynolds number and a calibration would be necessary for any particular valve; however, a figure of 0.62 could well be used as an initial value.

With hydraulic control valves it is necessary to relate the flow to the pressure drop from inlet to outlet of the valve i.e. points 1 and 3 in fig. 3.1. Let

p_3 = pressure at point 3
a_3 = area at point
v_3 = mean velocity of flow at point 3

Now the losses between points 2 and 3 may be considered to be those related to a sudden enlargement of section from a_2 to a_3. These losses may be expressed as

79

$$\text{pressure loss} = \frac{(v_2 - v_3)^2}{2}$$

Applying the steady-flow energy equation between points 1 and 3 in fig. 3.1 gives

$$\frac{p_1}{\rho} + \frac{v_1{}^2}{2} = \frac{p_3}{\rho} + \frac{v_3{}^2}{2} + \frac{\Delta p_f}{\rho} + \frac{(v_2 - v_3)^2}{2}$$

Assuming $v_1 = v_3$ then

$$\frac{1}{2} v_2{}^2 \left(1 - \frac{a_2}{a_3}\right)^2 = \frac{p_1 - p_3 - \Delta p_f}{\rho}$$

Therefore

$$v_2 = \left[1 - C_1 \frac{a_o}{a_1}\right]^{-1} \left[\frac{2}{\rho}(p_1 - p_3)\left(1 - \frac{\Delta p_f}{p_1 - p_3}\right)\right]^{\frac{1}{2}}$$

and

$$Q = \frac{C_1 \left[1 - \frac{\Delta p_f}{p_1 - p_3}\right]^{\frac{1}{2}}}{1 - C_1 \left[\frac{a_o}{a_1}\right]} \times a_o \left[\frac{2}{\rho}(p_1 - p_3)\right]^{\frac{1}{2}}$$

and this may be expressed as

$$Q = C_d' a_o \sqrt{\left[\frac{2(p_1 - p_3)}{\rho}\right]} \qquad (3.2)$$

where C_d' is now a discharge factor which permits the total valve pressure drop to be employed instead of the orifice pressure drop.

It should be noted that whereas C_d is not affected by cavitation beyond the valve orifice the value of C_d' will be affected by cavitation and a modified figure will have to be used once cavitation has started. C_d' may be taken as 0.75 for an average figure without cavitation.

Application to Control Valves

Speed-control Valve

This is a unidirectional variable-area orifice used to control the flow rate of the fluid. It is often fitted

with a full flow reverse path provided by a non-return
valve; see fig. 3.2.

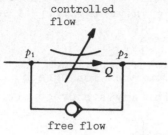

controlled
flow

p_1 p_2

Q

free flow

Fig.3.2

For this valve the flow rate is given by eqn 3.2 but p_2
now replaces p_3 as a symbol for downstream pressure
and $\Delta p_v = p_1 - p_2$. Therefore

$$Q = C_d^! a_o \sqrt{\left[\frac{2\Delta p_v}{\rho}\right]} \qquad\qquad (3.3)$$

In eqn 3.3, Q, Δp_v and a_o are all variables. Δp_v is determined by supply pressure and load pressure hence any
variation of load will vary the flow rate even if the
orifice area is unaltered. To obtain a linear relationship between Q and a_o it is necessary to keep Δp_v constant. This can be done by either series or parallel
pressure compensation.

Series Pressure Compensation

Fig. 3.3a shows a pressure compensating valve in series
with a variable-area orifice. Fig. 3.3b shows the
British Standard symbolic representation of this arrangement.

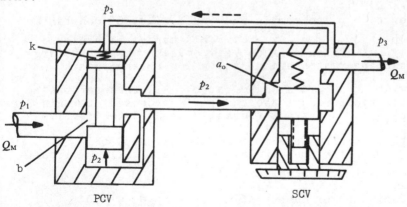

Fig.3.3a

In fig. 3.3a the variable orifice of the speed-control valve (S.C.V.) is a_o and is set manually. The spool of the pressure compensating valve (P.C.V.) is subjected to forces at both ends. Let

    A = P.C.V. spool area
    k = P.C.V. spring stiffness
    b = P.C.V. variable-area orifice
    x_0 = P.C.V. spring compression when b is closed
    x = opening of b for any steady condition Q_m,
p_2, p_3

Then $p_2 A = p_3 A + (- x + x_0) k$

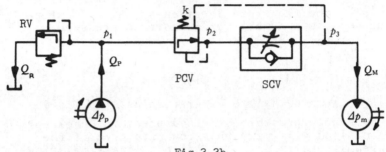

Fig.3.3b

Now the object is to maintain Q_m constant regardless of variations in Δp_m (= p_3 if tank gauge pressure is zero) Let there be a change in p_3 of δp_3 resulting in changes δp_2 and δx, therefore

$$(p_2 + \delta p_2) A = (p_3 + \delta p_3) A + (- x + x_0 - \delta x) k$$

and $\delta p_2 A = \delta p_3 A - \delta x\, k$

But for $p_2 - p_3$ to remain unchanged (i.e. Q_m constant) then $\delta p_2 = \delta p_3$, i.e. $\delta x = 0$. This cannot be achieved exactly but δx can be made very small by having a large compensating piston area A and a low spring rate k.

Port width b should also be large. All three of these specifications reduce the change δx necessary to maintain $p_2 - p_3$ = constant (normally about 3 to 4 bar).

With this arrangement it should be noted that if $Q_m < Q_p$ then the pressure p_1 is in fact maximum circuit pressure set by the relief valve.

    Useful power output at motor = $\Delta p_m Q_m = p_3 Q_m$

    $p_2 = p_3 + 3$ bar

    p_1 = relief valve pressure (bar)

Therefore power dissipated across orifice $a_o = 3 \times 10^5 Q_m$

power dissipated across orifice $b = \left[p_1 - (p_3 + 3 \times 10^5) \right] Q_m$

power dissipated across relief valve $= p_1 (Q_p - Q_m)$

Circuit efficiency $= \dfrac{p_3 Q_m}{p_1 Q_p}$

Hence low motor-speed and low motor-torque gives a very inefficient system with excessive oil heating and power wastage. This arrangement should only be used with low-power-level systems.

Parallel Pressure Compensation

In order to improve the system efficiency the compensating valve is placed in parallel with the variable orifice as shown in fig. 3.4a. Fig. 3.4b is the British Standard symbolic form for the arrangement.

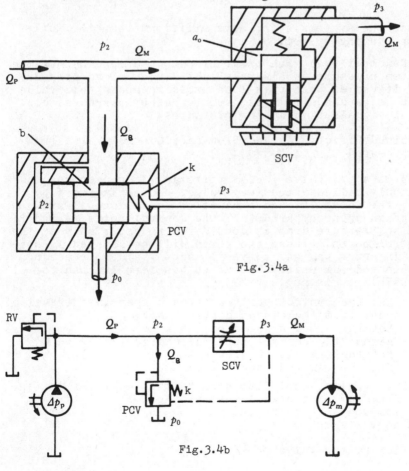

Fig.3.4a

Fig.3.4b

83

In this arrangement $p_2 - p_3$ is maintained constant by adjusting p_2 relative to p_0 (tank pressure) instead of p_1 (supply pressure). Since oil is passing to tank through the P.C.V. (Q_B) in parallel with the flow Q_m through the speed-control valve then p_2 bar becomes the maximum circuit pressure at any instant, i.e. pump pressure $(\Delta p_p) = (\Delta p_m + 3)$ bar at all flow rates Q_m except $Q_m = 0$ at which value $Q_B = Q_p$. At any flow Q_m (>0)

power dissipated across orifice $a_o = 3 \times 10^5 Q_m$

power dissipated across orifice $b = (p_3 + 3 \times 10^5)$ $(Q_p - Q_m)$

power dissipated across relief valve = 0

$$\text{Circuit efficiency} = \frac{p_3 Q_m}{(p_3 + 3 \times 10^5) Q_p}$$

Hence the system is far more efficient than the series compensating system.

Remember that with parallel compensation the maximum system pressure is always slightly above load pressure and it varies with this load pressure hence this value must be watched if it is used for other purposes, e.g. pilot signals or clamping pressures.

Combined Directional and Flow-rate Control Valve (non compensated)

This is a 5/3 valve i.e. 5 ports and 3 connections, commonly called a four-way valve. Referring to fig. 3.5, the inlet (P) and exhaust (T) ports are always large in area compared with the control ports (A and B) and so pressure drop is considered to occur only at the control ports. These two ports will be assumed identical in every way and they will supply a double-acting double rod ram unit. The valve has zero lap and compressibility is neglected.

Let the control ports area be b per unit travel of the spool from the closed centre position
Let the spool displacement = x
Assume the control ports discharge factor is C_d'
Let the ram piston area = A
Let the ram rod area = a

With the valve spool displaced to the right (as in fig. 3.5) the ram piston will move to the left relative to the ram body. Let the ram unit be loaded by a force F and have a speed v.

Flow at the valve ports

$$Q = C_d' bx \sqrt{\left[\frac{2(p_1 - p_2)}{\rho}\right]}$$

$$Q = C_d' bx \sqrt{\left[\frac{2(p_3 - p_4)}{\rho}\right]}$$

Thus $p_1 - p_2 = p_3 - p_4 = \Delta p_0$ (say)

Flow at the ram

$$Q = (A - a)v$$

Force balance at the ram

$$(p_2 - p_3)(A - a) = F$$

Now, total system pressure drop $= p_1 - p_4$

and $p_1 - p_4 = 2\Delta p_0 + (p_2 - p_3) = 2\Delta p_0 + \dfrac{F}{A - a}$

thus $2\Delta p_0 = (p_1 - p_4) - \dfrac{F}{A - a}$

and $C_d' bx \sqrt{\left[\dfrac{1}{\rho}(p_1 - p_4 - \dfrac{F}{A - a})\right]} = (A - a)v$ $\qquad$ (3.4)

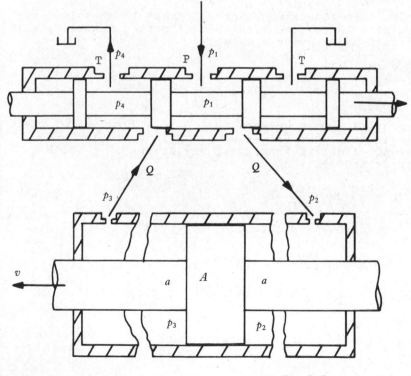

Fig.3.5

85

If the ram piston displacement relative to the body is y from the centre position then v = dy/dt. If p_1 = constant and F = constant then

$$\tau \frac{dy}{dt} = x \qquad (3.5)$$

where

$$\tau = \frac{A - a}{C_d' b \left[\frac{1}{\rho} \left(p_1 - p_4 - \frac{F}{A - a} \right) \right]^{\frac{1}{2}}} \qquad (3.6)$$

If p = constant but F is an inertia load, consisting of an equivalent mass M at the ram unit, then

$$F = M \frac{dv}{dt}$$

Substituting this into eqn 3.4 gives

$$C_d' bx \left[\frac{1}{\rho} \left(p_1 - p_4 - \frac{M dv/dt}{A - a} \right) \right]^{\frac{1}{2}} = (A - a)v$$

This is a non-linear equation which has to be solved by a special method. One such special method assumes small perturbations about mean values.

Now Q = (A - a)v $\qquad (3.7)$

and $Q = C_d' bx \sqrt{\left[\frac{1}{\rho} \left(p_1 - p_4 - \frac{F}{A - a} \right) \right]}$ $\qquad (3.8)$

From eqn 3.8 it can be seen that Q is a function of the two variables x and F. Assume a small variation of Q namely δQ resulting from small variations of x and F namely δx and δF, then

$$\delta Q = \frac{\partial Q}{\partial x} \delta x + \frac{\partial Q}{\partial F} \delta F \qquad (3.9)$$

Substituting eqn 3.9 into eqn 3.7 gives

$$\delta v = \frac{1}{A - a} \left(\frac{\partial Q}{\partial x} \delta x + \frac{\partial Q}{\partial F} \delta F \right)$$

Now $\partial Q / \partial x = K_x$ = valve gain i.e. the slope of the Q/x graph.

Also $\frac{F}{A - a} = (p_1 - p_4) - 2 \Delta p_0$

Thus $\frac{\partial Q}{\partial F} = \frac{\partial Q}{\partial \Delta p_0} \frac{\partial \Delta p_0}{\partial F} = \frac{\partial Q}{\partial \Delta p_0} \left(- \frac{1}{2(A - a)} \right)$

86

$$= - \frac{1}{2(A - a)} K_p$$

where $|Kp|$ = the slope of the $Q/\Delta p_0$ graph. Therefore

$$\delta v = \frac{1}{A - a} \left[K_x \, \delta x - \frac{Kp}{2(A - a)} \, \delta F \right]$$

For a pure inertia load $F = MDv$, therefore

$$\delta F = MD(\delta v)$$

thus $\delta v = \dfrac{K_x}{A - a} \, \delta x - \dfrac{MK_p}{2(A - a)^2} D(\delta v)$

and $\left[\dfrac{MK_p}{2(A - a)^2} D + 1 \right] \delta v = \dfrac{K_x}{A - a} \, \delta x$ $\qquad$ (3.10)

Eqn 3.10 will give the ram speed change δv as a result of a small spool position change δx. The solution of eqn 3.10 for a step change from $\delta x = 0$ to $\delta x = \delta x$ is

$$\delta v = \frac{K_x}{A - a} \, \delta x \left[1 - \exp\left(- \frac{t}{\tau} \right) \right]$$

where $\tau = MK_p/2(A - a)^2$, the system time-constant and $K_x/(A - a)$ is called the actuator steady-state gain.

Now, by definition, $K_p = \partial Q/\partial \Delta p_0$, and, since $Q = C_d' bx \sqrt{(2\Delta p_0/\rho)}$ then

$$\frac{\partial Q}{\partial \Delta p_0} = \frac{1}{2} C_d' bx \sqrt{\left(\frac{2}{\rho} \frac{1}{\Delta p_0} \right)}$$

thus $\tau = \dfrac{MC_d' bx}{4(A - a)^2} \sqrt{\left(\dfrac{2}{\rho \Delta p_0} \right)} = \dfrac{MQ}{4(A - a)^2 \Delta p_0} = \dfrac{Mv}{4(A - a)\Delta p_0}$

Note: This solution for a step change from $\delta x = 0$ to $\delta x = \delta x$ applies at any point where x has an initial value except $x = 0$. At $x = 0$, K_p and K_x cannot be computed and a computer solution of the original non-linear equation would be necessary. This is illustrated in one of the further examples at the end of the section.

WORKED EXAMPLES

1. Fig. 3.6 shows a three-way control valve connected to a single rod, double-acting, cylinder. The control

port of the valve is rectangular of width b and discharge factor C_d'. The supply pressure p_1 is constant.

When the piston velocity is v_1 to the right the load force is F_1 to the left and when the velocity is v_2 to the left the load is F_2 to the right.

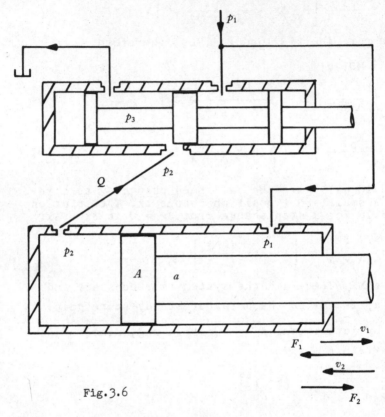

Fig.3.6

Establish expressions for v_1 and v_2 for equal displacements of the valve spool on either side of the zero lap centre position. Establish conditions by which v_1 and v_2 can be made equal in magnitude.

For v_1 let the valve spool move x to the left to produce the piston velocity v_1 to the right.

$$Q = Av_1 = C_d'bx \sqrt{\left[\frac{2(p_1 - p_2)}{\rho}\right]}$$

and $p_2A - p_1(A - a) = F_1$

thus $v_1 = \dfrac{C_d'bx}{A} \sqrt{\left\{\dfrac{2}{\rho}\left[p_1 - \dfrac{F_1}{A} - p_1\left(1 - \dfrac{a}{A}\right)\right]\right\}}$

88

i.e. $v_1 = \dfrac{C_d' bx}{A} \sqrt{\left[\dfrac{2}{\rho}\left(p_1\dfrac{a}{A} - \dfrac{F_1}{A}\right)\right]}$ (3.11)

For v_2 let the valve spool be displaced x to the right.

$$Q = Av_2 = C_d' bx \sqrt{\left[\dfrac{2(p_2 - p_3)}{\rho}\right]}$$

and $\quad p_1(A - a) - p_2 A = F_2$

thus $v_2 = \dfrac{C_d' bx}{A} \sqrt{\left\{\dfrac{2}{\rho}\left[- \dfrac{F_2}{A} + p_1\left(1 - \dfrac{a}{A}\right) - p_3\right]\right\}}$

i.e. $v_2 = \dfrac{C_d' bx}{A} \sqrt{\left\{\dfrac{2}{\rho}\left[p_1\left(1 - \dfrac{a}{A}\right) - \left(p_3 + \dfrac{F_2}{A}\right)\right]\right\}}$ (3.12)

For $v_1 = v_2$ from eqns 3.11 and 3.12

$$p_1\dfrac{a}{A} - \dfrac{F_1}{A} = p_1\left(1 - \dfrac{a}{A}\right) - \left(p_3 + \dfrac{F_2}{A}\right)$$

If $p_3 = 0$ (gauge) and if $F_1 = F_2$ then $v_1 = v_2$ provided $a/A = 1/2$ i.e. $A = 2a$. This condition is common in hydraulic equipment, hence $v_1 = v_2$ when $F_1 = F_2$ with $p_3 = 0$.

If $F_1 > F_2$ let the condition that $A = 2a$ be retained. Then

$$\tfrac{1}{2} p_1 - \dfrac{F_1}{A} = \tfrac{1}{2} p_1 - \left(p_3 + \dfrac{F_2}{A}\right)$$

If $v_1 = v_2$ then

$$\dfrac{F_1}{A} = p_3 + \dfrac{F_2}{A}$$ (3.13)

But if $F_1 > F_2$ then $p_3 A$ must provide the difference between F_1 and F_2; i.e. with $F_1 > F_2$, $v_1 = v_2$ provided p_3 is suitably adjusted by inserting a restrictor in the tank line from the valve.

If $F_1 < F_2$ from eqn 3.13 it would appear that p_3 would need to be negative. This is not a practical condition. However, if a restrictor is introduced between the supply line and the inlet to the three-way valve then the flow eqn 3.11 becomes

$$v_1 = \dfrac{C_d' bx}{A} \sqrt{\left\{\dfrac{2}{\rho}\left[(p_1 - \Delta p)\left(\dfrac{a}{A}\right) - \dfrac{F_1}{A}\right]\right\}}$$

89

and for $v_1 = v_2$

$$(p_1 - \Delta p)\left[\frac{a}{A}\right] - \frac{F_1}{A} = p_1\left[1 - \frac{a}{A}\right] - \left[p_3 + \frac{F_2}{A}\right]$$

Thus eqn 3.13 becomes

$$\frac{\Delta p}{2} + \frac{F_1}{A} = p_3 + \frac{F_2}{A}$$

Hence, with $F_1 < F_2$, Δp must be adjusted so that

$$\Delta p = 2\left[\frac{F_2 - F_1}{A} + p_3\right]$$

Fig. 3.7 shows the resulting circuit where complete control of v_1 and v_2 is possible for various values of F_1 and F_2.

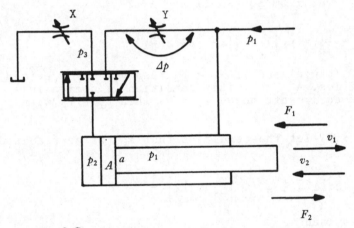

Fig.3.7

2. Fig. 3.8 shows a simple motor-speed-control system. Each control port has an area of 65 mm^2 and a discharge factor of 0.64. The fixed-capacity hydraulic motor has the following details

    Capacity = 32 cm^3/rev
    Volumetric efficiency = 88%
    Mechanical efficiency = 84%
    Speed = 16.7 rev/s
    Load at 16.7 rev/s = 2 kW

If the load pressure p_4 is 6×10^4 N/m^2 gauge and the oil density is 870 kg/m^3 determine the supply pressure p_1 to the valve.

Consider the valve to be in the left-hand position

At port B

$$Q_B = C_d' a \sqrt{\left(\frac{2\Delta p_B}{\rho}\right)}$$

At port A

$$Q_A = 0.88\ Q_B = C_d' a \sqrt{\left(\frac{2\Delta p_A}{\rho}\right)}$$

At the motor

$$\Delta p_m = \frac{2\pi\tau_m}{C_m\ \eta_{mech}} = \frac{2 \times 10^3}{16.7}\ \frac{1}{32 \times 10^{-6} \times 0.84} = 44.5\ \text{bar}$$

Also $Q_B = \dfrac{N_m C_m}{\eta_{vol}} = \dfrac{16.7 \times 32 \times 10^{-6}}{0.88} = 608 \times 10^{-6}\ \text{m}^3/\text{s}$

Hence

$$\frac{608 \times 10^{-6}}{65 \times 10^{-6} \times 0.64} = \sqrt{\left(\frac{2\Delta p_B}{870}\right)}$$

and $\dfrac{608 \times 10^{-6} \times 0.88}{65 \times 10^{-6} \times 0.64} = \sqrt{\left(\dfrac{2\Delta p_A}{870}\right)}$

thus $\Delta p_A = 0.716$ bar

$\Delta p_B = 0.930$ bar

Now $p_1 = p_4 + p_A + \Delta p_m + p_B$

$= 46.75$ bar

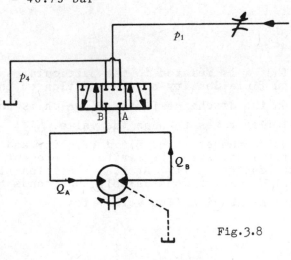

Fig.3.8

3. Show that the steady-state characteristics of a 4/2 hydraulic directional valve with no underlap may be represented by

$$\Delta p = \frac{Q^2}{K}$$

where Δp = pressure drop per control port
   Q = volumetric flow rate (equal for each control port)
   K = a constant for the valve

   Discuss how constant the value of K really is.

   A mass of 2500 kg is to be moved horizontally by a double-acting, through rod, cylinder unit controlled by a 4/2 valve. The effective piston area is 3×10^3 mm^2. Opposing the motion is a constant force of 500 N and a viscous resistance of 3 N s m^{-1}. A test on the valve reveals that it passes 500 ml/s when the pressure drop per control port is 25 bar.

   If the supply pressure is constant at 100 bar determine the velocity and acceleration of the load at maximum power output from the ram.

   The proof is given earlier in this section and results in eqn 3.2, i.e.

$$Q = C_d' a_0 \sqrt{\left(\frac{2\Delta p}{\rho}\right)}$$

Hence

$$\Delta p = \frac{Q^2}{K}$$

where

$$K = \frac{2}{\rho} \left(C_d' a_0\right)^2$$

It can be seen that K is related to the particular valve area a_0 and fluid density ρ. In addition to this K is dependent on the discharge factor C_d' which is not constant for all flow rates through the valve. C_d' varies with Reynolds number (i.e. fluid velocity and kinematic viscosity) and with the cavitation present downstream of the control port. At low velocities (low Reynolds numbers) C_d' varies considerably, but tends to become more constant at high flow velocities.

$$Q^2 = K\Delta p$$

thus $Q = \sqrt{(K\Delta p)}$

$$= Av = A \frac{dx}{dt}$$

i.e. $A \frac{dx}{dt} = \sqrt{(K\Delta p)}$ 　　　　　　　　　　　　　　　(3.14)

Also $(p_2 - p_3) A = m \frac{dv}{dt} + fv + F$ 　(see fig. 3.9)

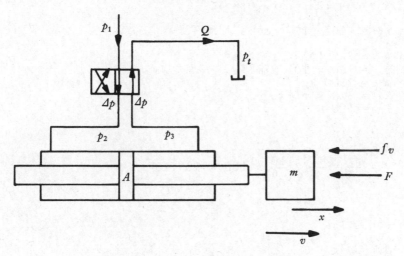

Fig.3.9

If $p_t = 0$ gauge, then

$$p_1 - 2\Delta p = p_2 - p_3$$

thus $A(p_1 - 2\Delta p) = m \frac{d^2x}{dt^2} + f \frac{dx}{dt} + F$ 　　　　　　　(3.15)

For maximum power-output at ram

$$2\Delta p = \frac{1}{3} (p_1 - p_t)$$

i.e. total losses $= \frac{1}{3}$ (available pressure)

Also, if $p_t = 0$, then

$$\Delta p = \frac{1}{6} p_1$$

thus eqn 3.14 gives

$$\frac{dx}{dt} = \frac{1}{A} \sqrt{\left(\frac{K}{6} p_1\right)}$$ 　　　　　　　　　(3.14a)

93

and eqn 3.15 gives

$$m \frac{d^2x}{dt^2} + f \frac{dx}{dt} + F = \frac{2A}{3} p_1 \qquad\qquad (3.15a)$$

For K, given that $Q = 500$ ml/s when $\Delta p = 25$ bar, then

$$(500 \times 10^{-6})^2 = K \times 25 \times 10^5$$

$$K = 1 \times 10^{-13} \text{ or } 10 \times 10^{-14}$$

thus $\sqrt{K} = \sqrt{10} \times 10^{-7} \ m^4 \ s^{-1} \ N^{-\frac{1}{2}}$

Also $A = 3 \times 10^3 \times 10^{-6} \ m^2$
$ p = 100 \times 10^5 \ N/m^2$
$ m = 2500 \ kg$
$ f = 3 \ N \ s \ m^{-1}$
$ F = 500 \ N$

From eqn 3.14a

$$v = \frac{dx}{dt} = \frac{10^3}{3} \sqrt{\left(\frac{10}{6} \times 100 \times 10^5 \right) 10^{-7}}$$

$$= 0.136 \ m/s$$

From eqn 3.15a

$$\frac{dv}{dt} = \frac{d^2x}{dt^2} = \frac{1}{m} \sqrt{\left(\frac{2A}{3} p_1 - f \frac{dx}{dt} - F \right)}$$

$$= \frac{1}{2.5 \times 10^3} \left(\frac{2}{3} \times 3 \times 10^4 - 3 \times 0.136 - 500 \right)$$

$$= 0.78 \ m/s^2$$

4. The pilot relief valve shown in fig. 3.10 has a main spool spring equivalent to a pressure differential k_1 across the spool at cracking (i.e. spool just opening port to tank). The pilot (poppet valve) spring is equivalent to a pressure k_2 in chamber B when just open to tank (nominal relief pressure). The chamber B has a volume V_B of oil of effective bulk modulus B. The normal system pressure is p_N and the valve is closed.

Ignore spool mass and viscous damping. Show that the valve will respond to a rate of change of pressure p in chamber A, where $p > p_N$, due to some increase of system load. Assume the rate of change to be constant i.e. $dp/dt = R$.

Obtain an expression for the response time of the valve for various values of R and determine the minimum value of R for which the valve will respond to a rate of change.

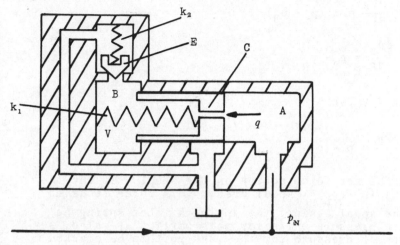

Fig.3.10

What will be the minimum response time of the valve when R is too small to cause a response to rate of change?

Consider the system pressure to have a normal value of p_N. The pressure in chambers A and B will also be p_N. Let the pressure p_N now increase due to a valve closure or a high inertia load. Let the new value of p_N be p at any instant. Let the pressure in chamber B be p_B when the new pressure in chamber A is p (p > p_B > p_N). Let the flow through the orifice in spool C be laminar and consider this as q = α (p - p_B) where α is a constant.

Provided poppet valve E has not opened, the flow q is due entirely to the compressibility of the oil in chamber B. By definition

$$B = - V_B \frac{dp_B}{dV_B}$$

Also $q = - \frac{dV_B}{dt}$

i.e. q is positive when oil flows into chamber B and the original volume of oil V_B is being compressed. Therefore

95

$$q = -\frac{dV_B}{dt} = +\frac{V_B}{B}\frac{dp_B}{dt}$$

and $\dfrac{V_B}{B}\dfrac{dp_B}{dt} = \alpha\,(p - p_B)$

i.e. $\tau\,\dfrac{dp_B}{dt} = p - p_B$

where $\tau = V_B/\alpha B$

$$\tau\,\frac{dp_B}{dt} = p - p_B \tag{3.16}$$

Now the net force acting on the spool C at any instant is $(p - p_B) \times$ spool area. This force tends to move the spool against the spring k. Let spring be rated in pressure units for this particular valve, say k_1 units of pressure for the tank port to be cracked open. Similarly let poppet valve spring be rated at k_2 pressure units.

Let $p - p_B = \Delta p$ \hfill (3.17)

Consider the pressure p to be

$$p_N + Rt \tag{3.18}$$

where R is a constant rate of pressure rise

i.e. $R = \dfrac{dp}{dt}$ \hfill (3.19)

Substituting eqn 3.17 into eqn 3.16 gives

$$\tau\,\frac{d(p - \Delta p)}{dt} = \Delta p$$

thus $\tau\,\dfrac{d(\Delta p)}{dt} + \Delta p = \tau\,\dfrac{dp}{dt}$

and substituting eqn 3.19 into this equation gives

$$\tau\,\frac{d(\Delta p)}{dt} + \Delta p = \tau R$$

Now $\Delta p = 0$ at $t = 0$, therefore

$$\Delta p = \tau R \left[1 - \exp(-\,t/\tau) \right] \tag{3.20}$$

Considering eqn 3.16 again gives

$$\tau \frac{dp_B}{dt} + p_B = p$$

Now $p_B = p_N$ at $t = 0$ and $p = p_N$ at $t = 0$, therefore

$$\tau \frac{dp_B}{dt} + \Delta p_B = \Delta p_N = Rt$$

where $\Delta p_B = p_B - p_N$, $\Delta p_N = p - p_N$ and $dp_B/dt = d(\Delta p_B)/dt$, therefore

$$\Delta p_B = Rt - \tau R \left[1 - \exp(- t/\tau)\right] \qquad (3.21)$$

(i.e. $\Delta p_B = \Delta p_N - \Delta p$.)

The spool C will open when $p - p_B > k_1$, i.e. crack at $\Delta p = k_1$, therefore

$$k_1 = \tau R \left[1 - \exp(- t/\tau)\right]$$

Thus response time of valve $= t = \tau \log_e \left[1/(1 - \frac{k_1}{\tau R})\right]$

This may be calculated for various rates of pressure rise R and plotted as in fig. 3.11 as t/τ v. $R/(k_1/\tau)$ During this time the system pressure will have risen from p_N to $p_{max} = p_N + Rt$.

Fig. 3.11 also shows $(p_{max} - p_N)/k_1$ v. $R/(k_1/\tau)$. From this it can be seen that as the rate of pressure rise increases (R increasing) the maximum system pressure reached actually decreases due to the faster response. Also the response time is infinite at $R/(k_1/\tau) = 1$, i.e. the valve will not respond to rates of pressure rise below $R = k_1/\tau$.

Fig. 3.12 shows the maximum system pressure plotted against system response time $(p_{max} - p_N)/k_1$ v. t/τ.

Fig. 3.13 shows the variation of Δp, p_{max} and p_B with time (plotted as dimensionless parameters). From this graph the values of maximum system pressure p_{max} and p_B can be obtained after any time interval up to the response time where $\Delta p = k_1$ and then $\Delta p/R\tau = k_1/R\tau$. Again, at $k_1/R\tau = 1$, $t/\tau = \infty$ and the valve does not respond to rates of change of pressure. However, it will respond as a normal pilot relief valve.

97

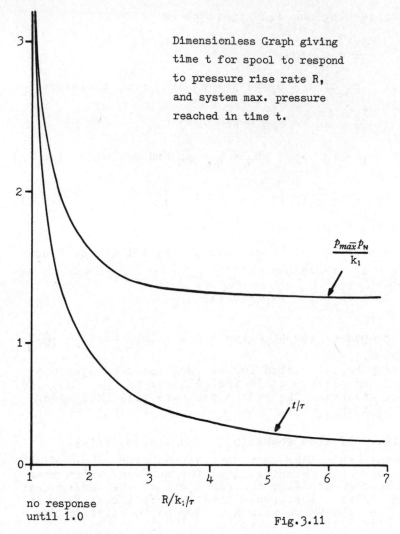

Dimensionless Graph giving
time t for spool to respond
to pressure rise rate R,
and system max. pressure
reached in time t.

$\dfrac{p_{m\bar{a}x}\,p_N}{k_1}$

t/τ

no response
until 1.0

$R/k_:/\tau$

Fig.3.11

Now $p_B = p_N + \Delta p_B$

and using eqn 3.21

$$p_B - p_N = Rt - \tau R\left[1 - \exp(-\,t/\tau)\right]$$

$$= Rt - \Delta p$$

i.e. $\dfrac{p_B - p_N}{\tau R} = \dfrac{t}{\tau} - \dfrac{\Delta p}{\tau R}$

Since the spool does not respond to rate of change
of pressure, then

98

$$\frac{\Delta p}{\tau R} = \frac{k_1}{\tau R} = 1$$

(at maximum R for no rate response) and time for p_B to rise to spring value k_2 is given by

$$\frac{k_2 - p_N}{\tau R} = \frac{t}{\tau} - 1$$

i.e. $\frac{t}{\tau} = 1 + \frac{k_2}{k_1} - \frac{p_N}{k_1}$ (3.22)

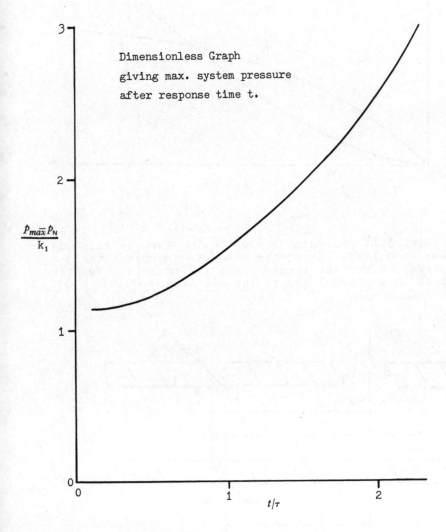

Dimensionless Graph
giving max. system pressure
after response time t.

$\frac{p_{max} - p_N}{k_1}$

t/τ

Fig.3.12

99

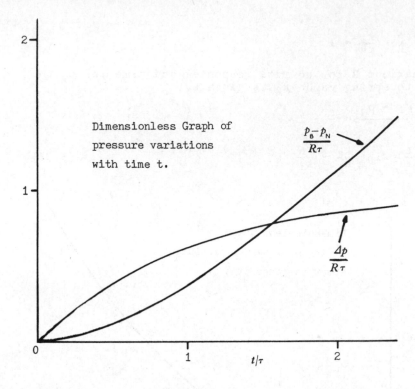

Dimensionless Graph of
pressure variations
with time t.

$\frac{p_B - p_N}{R\tau}$

$\frac{\Delta p}{R\tau}$

Fig.3.13

5. Fig. 3.14 indicates a spool valve through which oil
flows to a load. Using the standard symbols establish
expressions for the steady state and transient reaction
forces on the spool due to the oil flow.

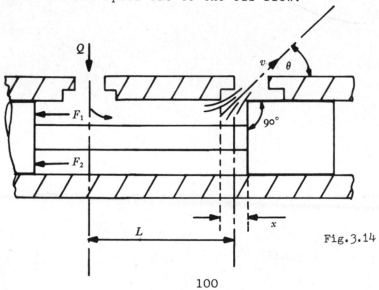

Fig.3.14

100

Establish an equation that represents the system and discuss the effect of introducing a second valve chamber as indicated in fig. 3.15.

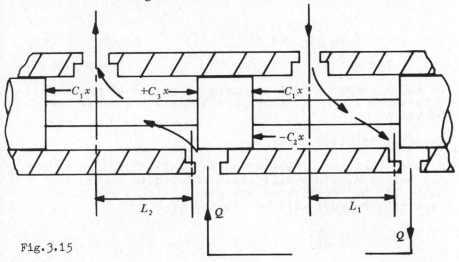

Fig.3.15

Steady Reaction Force

Consider the steady flow of the oil through the valve as indicated in fig. 3.14. It is assumed that the inlet port opening is large, compared with the control orifice, so that inlet velocity can be ignored. The flow is also assumed to be ideal steady two-dimensional flow for an ideal fluid. The flow and velocity are assumed to be positive when directed out of the valve chamber.

The rate of change of momentum of the oil through the valve chamber is

$$\frac{d(mv)}{dt} = v \frac{dm}{dt} = \rho vQ$$

where ρ = density
      Q = volumetric flow rate

The force to produce the component of this rate of change, parallel to the spool axis, will come from the spool and so this may be considered to be subjected to a force, from the fluid, in the opposite direction (F_1).

$$F_1 = -\rho vQ \cos \theta$$

Now for a constant control port pressure drop Δp_0, discharge coefficient C_d and area a_0 per unit port opening x

101

$$F_1 = - \rho C_d a_0 x \frac{2 \Delta p_0}{\rho} \cos \theta$$

$$= - C_1 x$$

where $C_1 = 2 C_d a_0 \Delta p_0 \cos \theta$

i.e. $F_1 = - C_1 x$ (3.23)

The radial component of the rate of change of momentum is balanced if the control port is symmetrical around the valve.

Transient Reaction Force

Consider the valve spool to be suddenly displaced to the open position. This will cause the liquid within the chamber to be accelerated and an additional reaction force on the spool will be created (F_2)

$$F_2 = - \rho A L \frac{dv}{dt}$$

where A = cross section area of valve chamber.

Now $Q = Av$

Thus $\frac{dQ}{dt} = A \frac{dv}{dt}$

and $F_2 = - \rho L \frac{dQ}{dt}$

$$= - \rho L \frac{d}{dt} \left[C_d a_0 x \sqrt{\left(\frac{2 \Delta p_0}{\rho} \right)} \right]$$

If the control port pressure drop (Δp_0) is assumed to be constant then

$$F_2 = - \rho L C_d a_0 \sqrt{\left(\frac{2 \Delta p_0}{\rho} \right)} \frac{dx}{dt}$$

$$= - C_2 \frac{dx}{dt}$$

where $C_2 = C_d a_0 L \sqrt{(2 \rho \Delta p_0)}$

i.e. $F_2 = - C_2 \frac{dx}{dt}$ (3.24)

If the valve spool has a mass M and is subject to a viscous damping force fv and a spring force kx then

$$M \frac{d^2 x}{dt^2} + f \frac{dx}{dt} + kx = F_1 + F_2$$

102

thus $M \dfrac{d^2x}{dt^2} + (f + C_2) \dfrac{dx}{dt} + (k + C_1) x = 0$ (3.25)

Consider now fig. 3.15. An additional control port and valve chamber have been added to the original fig. The flow rate and velocity in this second chamber are both negative (both directed into the chamber) and the length L_2 is negative for L_1 positive. Hence

total force $F_1 = - 2 C_1 x$

and $F_2 = (- C_2 + C_3) \dfrac{dx}{dt}$

where

$$C_2 = C_d a_0 \, L_1 \sqrt{(2 \rho \Delta p_0)}$$

and $C_3 = C_d a_0 \, L_2 \sqrt{(2 \rho \Delta p_0)}$

Hence eqn 3.25 is modified to

$$M \dfrac{d^2x}{dt^2} + (f + C_2 - C_3) \dfrac{dx}{dt} + (k + 2C_1) x = 0 \quad (3.26)$$

From eqn 3.26 the undamped natural frequency of the spool is

$$\omega_n = \sqrt{\left(\dfrac{k + 2C_1}{M} \right)} \qquad (3.27)$$

and the damping ratio ξ is obtained from

$$2 \xi \omega_n = \dfrac{f + C_2 - C_3}{M}$$

$$\xi = \dfrac{f + C_2 - C_3}{2\sqrt{[M(k + 2C_1)]}} \qquad (3.28)$$

also, the damped natural frequency is

$$\omega = \omega_n \sqrt{(1 - \xi^2)} \qquad (3.29)$$

If $0 < (f + C_2 - C_3) < 1.0$ the spool will oscillate and finally settle into a steady-state position where $(k + 2C_1) x = 0$ i.e. the spool is held open by the spring force.

If, in eqn 3.26 $(f + C_2 - C_3) < 0$

i.e. $C_3 > C_2 + f$ (3.30)

then the valve spool will oscillate continuously i.e. it has negative damping.

Eqn 3.30 yields

$$C_d a_0 \ L_2 \sqrt{(2\rho \Delta p_0)} > C_d a_0 \ L_1 \sqrt{(2\rho \Delta p_0)} + f$$

for oscillation.

If the viscous friction on the spool can be neglected initially then it can be deduced that the condition for stability is that $L_1 > L_2$, i.e. the valve chamber volume associated with positive damping (control port at outlet) is greater than that associated with negative damping (control port at inlet).

6. Fig. 3.16 indicates a double-acting through rod actuator supplied by a 4/3 metering valve for which the following data apply.

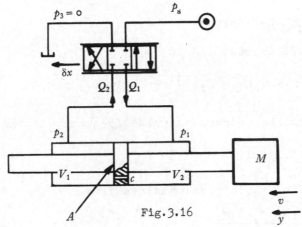

Fig.3.16

Supply pressure p_s = constant
Linearised valve flow $\delta Q = K_x \delta x + K_p \delta (\Delta p)$ where Δp is the control orifice pressure drop
Leakage flow through orifice c = $c(p_1 - p_2)$
Bulk modulus = B
Actuator load = mass M
Total oil volume = V

Determine the transfer function relating

(a) piston speed to valve spool displacement ($\delta v / \delta x$)
(b) piston position to valve spool displacement
    ($\delta y / \delta x$)
Linearisation

If $Q = kx \sqrt{(p_s - p_1)}$

then $\delta Q = \delta x \left[k \sqrt{(p_s - p_1)} \right] - \dfrac{\delta p_1}{2} \left[\dfrac{kx}{\sqrt{(p_s - p_1)}} \right]$

104

thus $\dfrac{\delta Q}{Q} = \dfrac{\delta x}{x} - \dfrac{\delta p_1}{2}\,\dfrac{1}{(p_s - p_1)}$

i.e. $\delta Q = \left(\dfrac{Q}{x}\right)\delta x - \left[\dfrac{Q}{2(p_s - p_1)}\right]\delta p_1$

thus $\delta Q = K_x\,\delta x + K_p\,\delta p_1$

where

$$K_x = \dfrac{\partial Q}{\partial x} = \dfrac{Q}{x}$$

and $K_p = \dfrac{\partial Q}{\partial p_1} = -\dfrac{Q}{2(p_s - p_1)}$

$$Q_1 = K_x\,\delta x - K_p\,\delta p_1 = A\,\delta v + c(\delta p_1 - \delta p_2) + \dfrac{V_1}{B}s(\delta p_1)$$

$$Q_2 = K_x\,\delta x + K_p\,\delta p_2 = A\,\delta v + c(\delta p_1 - \delta p_2) - \dfrac{V_2}{B}s(\delta p_2)$$

Therefore $\hspace{3cm}$ (s = Laplace operator)

$$K_x\,\delta x + \dfrac{K_p}{2}(\delta p_2 - \delta p_1) = A\,\delta v + c(\delta p_1 - \delta p_2) +$$
$$\dfrac{1}{2B}\,s\,(V_1\delta p_1 - V_2\delta p_2)$$

For minimum natural frequency $V_1 = V_2$ and $V_1 + V_2 = V$.
Also

$$Ms(\delta v) = A(\delta p_1 - \delta p_2)$$

thus $K_x\,\delta x - \dfrac{K_p}{2}\dfrac{M}{A}s(\delta v) = A\,\delta v + c\dfrac{M}{A}s(\delta v) + M\dfrac{V}{4BA}s^2(\delta v)$

and $\left[\dfrac{MV}{4BA^2}s^2 + \left[\dfrac{MK_p}{2A^2} + \dfrac{cM}{A^2}\right]s + 1\right]\delta v = \dfrac{K_x}{A}\,\delta x$

thus $\dfrac{\delta v}{\delta x} = \dfrac{K_x/A}{\dfrac{1}{\omega_n^{\,2}}s^2 + \dfrac{2\xi}{\omega_n}s + 1}$

where

$$\omega_n^{\,2} = \dfrac{4BA^2}{MV}$$

and $\xi = \left(\dfrac{K_p}{2} + c\right)\sqrt{\left(\dfrac{MB}{VA^2}\right)}$

since $\delta v = s(\delta y)$, then

105

$$\frac{\delta y}{\delta x} = \frac{K_x/A}{s\left[\frac{s^2}{\omega_n^2} + \frac{2\xi}{\omega_n}s + 1\right]}$$

Note that K_x/A will be given the symbol K_L and, in the section dealing with servo systems, will be referred to as the steady-state gain of the actuator.

FURTHER EXAMPLES

1. Fig. 3.17 shows a 3/3 valve and actuator unit. The constant supply pressure is 100 bar and the tank pressure is 2 bar gauge. The valve control port area is 30 mm^2, the discharge factor is 0.7, the oil density is 882 kg/m^3 and A = 2a.

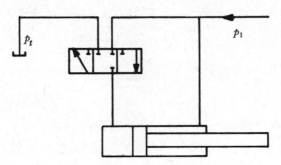

Fig.3.17

When the piston advances the load will be equivalent to 4 kN and when it retracts the load will be 800 N. If the piston speed is to be 1 m/s in both directions determine

(a) the area A
(b) the pressure drop across the restrictor that must be introduced into the circuit   (A = 10^3 mm^2 Δp = 30 bar)

2. Describe the action of the valve illustrated in fig. 3.18

Numerical Example

piston load F = 10^4 N when piston displacement is zero
piston load F = 5 × 10^3 N when piston displacement is full stroke
load/displacement graph is a straight line between these two extremes

106

piston effective area = 2×10^3 mm^2
orifice discharge factor C_d = 0.62

oil density = 870 kg/m^3
supply pressure is constant at 100 bar

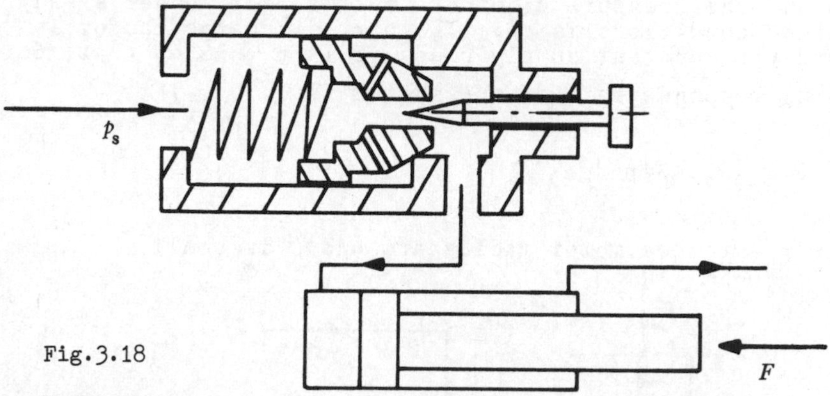

Fig.3.18

Determine

(a) the ratio of piston maximum speed to minimum speed during outstroke, assuming the valve needle to be unchanged

(b) the valve needle orifice area for a piston midstroke power output of 15 kW

(1.22:1, 54 mm^2)

3. A pilot relief valve of the type shown in fig. 3.10 has the following details

s_1 = 7 bar
s_2 = 70 bar
V_B = 30 ml
B = 1.7 GN/m^2
p_N = 35 bar
α = 35.2 $\times 10^{-3}$ ml/bar s

Determine the response time of the valve and the maximum system pressure when the steady rate of system pressure rise is (a) 400 bar/s, (b) 80 bar/s and (c) the greatest rate of rise that is just insufficient for the valve to respond to the rate of change.

[(a) 21.6 ms, 43.64 bar]
[(b) 525 ms, 77 bar]
[(c) 300 ms, 77 bar]

4. Electro-hydraulic servo valves frequently employ torque motors as an interface between the electrical error signal and the valve spool displacement. Such an arrangement is shown in fig. 3.19 where the valve is a 5/3 zero lap type as illustrated previously in fig. 3.5. The torque motor output torque may be expressed as T = ϕ(i) - aβ where ϕ(i) indicates some function of the error current i

107

a = torque/radian displacement of torque arm
β = torque arm displacement

The valve control port area per unit displacement of spool is b and the pressure drop per control port, under steady unloaded conditions is Δp. The port discharge factor is C_d and the constant supply pressure is p_s. Show that, for maximum response rate of the system (i.e. Q_{max})

$$\frac{a}{C_d b} = 1.434 r^2 \Delta p$$

where r = torque motor radius arm and β is small.

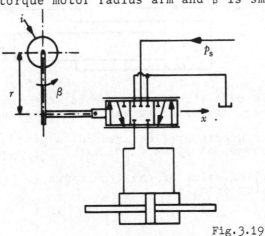

Fig. 3.19

5.   Fig. 3.20 shows a hydraulic drive where compressibility may be ignored. The mass of oil in the pipeline to the actuator is m_1 and the mass in the actuator is m_2. The piston area is A and the flow equation for valve B is $Q_B = k p_2$.

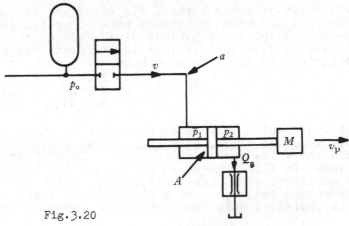

Fig. 3.20

108

Viscous friction in the pipeline is equivalent to a pressure loss fv where v is the oil velocity. Valve C may be opened to subject the system to a pressure p_0.

Show that the system response is represented by the equation

$$(\tau s + 1)v_p = \frac{A}{R} p_0$$

where $\tau = \frac{M^1}{R}$, v_p = piston speed

$$M^1 = \text{equivalent mass at piston} = M + m_2 + \frac{A^2}{a^2} m_1$$

$$R = A^2 \left[\frac{1}{k} + \frac{f}{a} \right]$$

6. Fig. 3.21 shows a flapper nozzle valve and a single acting hydraulic actuator. The supply pressure p_1 is constant and the flow through orifice is given by $Q_1 = K_1\sqrt{(p_1 - p_2)}$. The flow through the nozzle 2 is given by

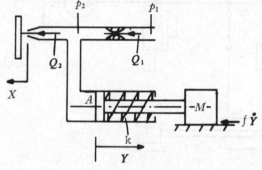

Fig.3.21

$Q_2 = K_2X\sqrt{p_2}$ where X is the flapper distance from the nozzle. The load on the actuator consists of the mass M and a viscous resistance of value f per unit piston velocity. The spring rate is k per unit deflection of the spring and the piston area is A.

Neglecting compressibility and using a linearisation method show that the transfer function relating small changes of Y (i.e. y) to small changes of X (i.e. x) may be expressed as

$$\frac{y}{x} = \frac{-\dfrac{C_3}{C_1 + C_2} \dfrac{A}{M}}{s^2 + \left[\dfrac{f}{M} + \dfrac{A^2}{M(C_1 + C_2)} \right] s + \dfrac{k}{M}}$$

where $C_1 = K_1/2\sqrt{(p_1 - p_2)}$
  $C_2 = K_2X/2\sqrt{p_2}$
  $C_3 = K_2\sqrt{p_2}$

7. Fig. 3.22 indicates a double acting, differential area actuator supplied via a 3/3 metering valve for which the following data apply

  p_s = constant
  linearised valve flow $\delta Q = K_x\delta x + K_p \delta(\Delta p)$ where Δp
is the control port pressure drop
  leakage flow through orifice $c = c(p_s - p_1)$

  bulk modulus = B
  actuator load = M (mass)
  volume of oil in line between valve and actuator head end = v

  Determine the transfer function relating piston movement to spool displacement for small movements $(\delta y/\delta x)$

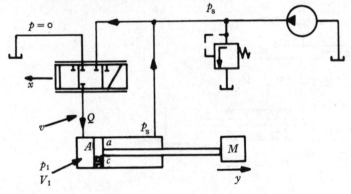

Fig. 3.22

$$\frac{\delta y}{\delta x} = \frac{K_x/A}{s\,(\tau^2 s^2 + 2\zeta\tau\,s + 1)}$$

$$\tau = \sqrt{\left(\frac{Mv}{A^2 B}\right)} \quad \text{for minimum natural frequency}$$

$$\zeta = \left(\frac{K_p + c}{2}\right)\sqrt{\left(\frac{BM}{A^2 v}\right)}$$

8. The flow through a hydraulic spool valve is given by the equation $Q = 60x\sqrt{(\Delta p)}$

where Q = ml/s

x = valve opening mm

Δp = total valve pressure drop (bar)

The valve is a symmetrical 5/3 one i.e. it has two control ports, and the constant supply pressure is 100 bar
with a tank port pressure of zero gauge.

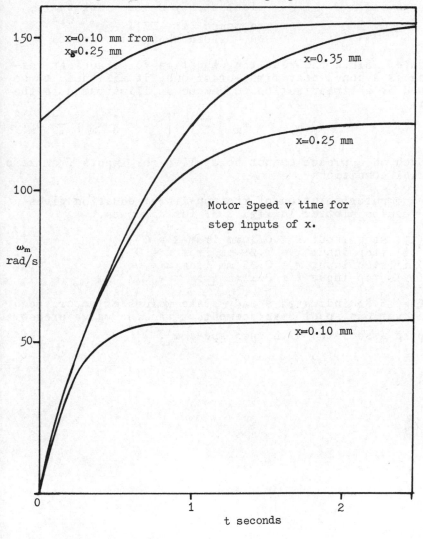

x=0.10 mm from
x=0.25 mm

x=0.35 mm

x=0.25 mm

Motor Speed v time for

step inputs of x.

x=0.10 mm

Fig.3.23

A motor with the following characteristics is connected
to the control ports of the valve

111

moment of inertia = 5×10^{-2} kg m^2
viscous friction = 0.03 N s m
capacity = 6 ml/rev

Determine the speed/time response of the motor when the valve spool is given a step displacement from x = 0.25 mm of x = 0.10 mm.

[initial steady speed = 125 rad/s,
final steady speed = 162 rad/s,
response $\delta\omega = 37(1 - e^{-t/0.4})$,
$\delta\omega$ = speed change in time t]

Note: Since the resulting equation for transient response is a non-linear first-order one, it will need to be solved by a linearisation technique as illustrated in the text.

$$\omega_m = 20\pi x \left[100 - \frac{\pi}{6} \dot{\omega}_m - \frac{\pi}{10} \omega_m\right]^{\frac{1}{2}}$$

Such an approach cannot be applied for inputs from zero initial conditions.

A computer solution of the non-linear equation gives the results plotted in fig. 3.23 for 4 inputs.

(i)   step input x = 0.1 mm from x = 0
(ii)   step input x = 0.25 mm from x = 0
(iii) step input x = 0.35 mm from x = 0
(iv)   step input x = 0.1 mm from x = 0.25 mm

Fig. 3.23 indicates steady-state values of motor speed ω_m at various spool displacements x and the valve pressure drop Δp associated with each speed ω_m.

4 ACCUMULATOR SYSTEMS

Accumulators form an energy store in a hydraulic system and have two common uses

(1) To provide a source of energy that may be required over a short period of time. In such a system a small pump (high pressure, small volumetric output) is commonly used to charge the accumulator over a relatively long time period and then a sudden demand for energy is made over a small time period; this being provided by the accumulator.
(2) To smooth out pressure fluctuations in a system e.g. variations in pump delivery and transient pressures due to sudden changes of fluid velocity in the system.

Accumulators in hydraulic systems are generally of the synthetic rubber bag type or the free piston type. Both types provide a boundary between the working hydraulic fluid (oil) and a compressed gas (inert) which stores the energy. The following mathematics applies to both types.

SIMPLE ANALYSIS

Flow Graph

In a system where flow demands on the pump are of a variable nature, a fixed delivery pump, capable of meeting the maximum demand would be wasteful of power and a smaller pump may be used in association with an accumulator. Fig. 4.1 illustrates a possible system flow graph where the flow rate is plotted against time. The average flow rate Q_A may be estimated from

$$Q_A = \frac{\sum Q \times t}{\sum t} \text{ for } t = 0 \text{ to } t = 25$$

Flows above the average are provided by the accumulator and flows below it are provided by the pump.

In addition to the volume of oil required from the accumulator it is necessary to know the minimum pressure at which this oil has to be delivered. When the accumulator is charged with oil the gas (in the bag or above the free piston) will be compressed and the pump must be capable of providing the maximum pressure of the gas when the accumulator is fully charged. For practical reasons it is best to operate an accumulator in such a way that it is never completely empty of oil (approximately 10% of the oil volume is kept in the accumulator).

113

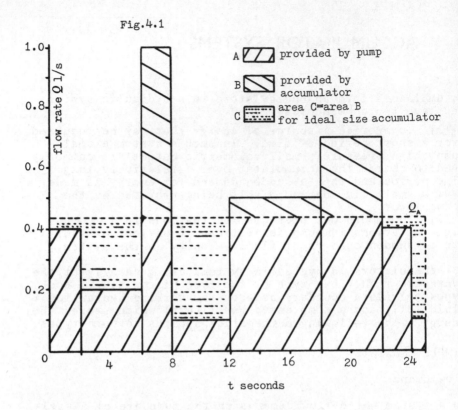

Fig.4.1

A ▨ provided by pump

B ◣ provided by accumulator

C ▦ area C=area B for ideal size accumulator

flow rate Q l/s

t seconds

Size of Accumulator

Fig. 4.2 represents the three main stages of charge of an accumulator. The moving line represents the boundary between the oil and gas (bag or piston).

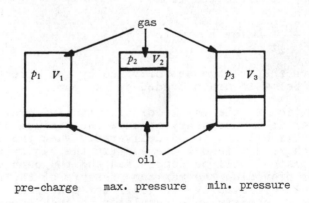

p_1 V_1 gas p_2 V_2 p_3 V_3

oil

pre-charge   max. pressure   min. pressure

Fig.4.2

Let p_1 = initial gas pressure (absolute).
p_2 = maximum gas pressure (absolute) when fully charged (i.e. gas volume is minimum).
p_3 = minimum pressure (absolute) at which the system can operate.
V_1 = initial volume of gas in the accumulator.
V_2 = volume of gas at maximum pressure.
V_3 = volume of gas at pressure p_3.

It should be noted that volumes V_1, V_2 and V_3 include the gas bottles associated with the accumulator.

The volume of oil delivered from the accumulator during the pressure drop p_2 - p_3 is V_3 - V_2. The physical size of the accumulator will be taken as V_1 when calculated, and then the nearest standard size accumulator greater than V_1 is selected since V_1 is the actual gas volume. Assuming a perfect gas and constant mass of gas

$$\frac{p_1 V_1}{T_1} = \frac{p_2 V_2}{T_2} = \frac{p_3 V_3}{T_3}$$

where T is the absolute temperature of the gas at each stage.

Therefore

$$\text{oil delivered} = V_3 - V_2 = V_1 \left(\frac{p_1}{p_3} \frac{T_3}{T_1} - \frac{p_1}{p_2} \frac{T_2}{T_1} \right)$$

and accumulator size $= V_1 = \frac{p_2}{p_1}(V_3 - V_2) \left/ \frac{T_2}{T_1} \left(\frac{p_2}{p_3} \frac{T_3}{T_2} - 1 \right) \right.$

$$(4.1)$$

If the charge and discharge processes are such that the gas remains at constant temperature, then

$$V_1 = \frac{p_2}{p_1}(V_3 - V_2) \left/ \left(\frac{p_2}{p_3} - 1 \right) \right. \qquad (4.2)$$

If the discharge process is very rapid, then

$$\frac{p_1 V_1}{T_1} = \frac{p_2 V_2}{T_2}$$

and $\quad p_2 V_2{}^\gamma = p_3 V_3{}^\gamma$

where γ = 1.4 (average). Therefore

$$V_1 = \frac{p_2}{p_1}(V_3 - V_2) \left/ \frac{T_2}{T_1} \left[\left(\frac{p_2}{p_3} \right)^{1/\gamma} - 1 \right] \right. \qquad (4.3)$$

and if $T_2 = T_1$ i.e. isothermal charge process

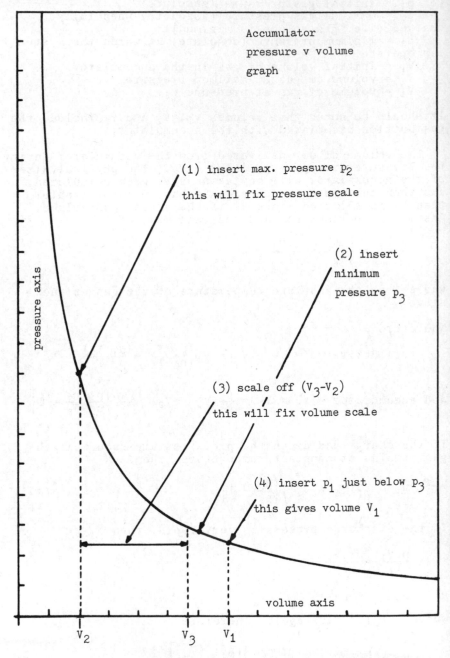

Fig.4.3

$$V_1 = \frac{p_2}{p_1}(V_3 - V_2) \bigg/ \left[\left(\frac{p_2}{p_3}\right)^{1/\gamma} - 1 \right] \tag{4.4}$$

Eqns 4.1, 4.2, 4.3 and 4.4 can be used to calculate the accumulator size under different conditions of operation. Because the expansion process may vary in rate it is common to use the law $pV^{1.2} = $ constant instead of the ideal $pV^\gamma = $ constant.

Fig. 4.3 is a plot of $pV = $ constant i.e. isothermal curve. This can be used to solve many simple accumulator problems as the worked examples illustrate. To use this curve

(1) first insert p_2; this will fix the pressure axis scale
(2) insert p_3
(3) scale off $(V_3 - V_2)$ between p_2 and p_3 or slightly higher than p_3; this will fix the volume axis scale
(4) select p_1 (pre-charge pressure) just below p_3 (say 10 bar below)
(5) measure off V_1 to scale determined by (3)

Fig. 4.4 shows $pV = $ constant plotted on a log-log graph giving a straight line on which calculations can be made. Included on the figure are lines for $pV^{1.4} = $ constant and $pV^{1.2} = $ constant so that these processes may be considered during the accumulator discharge period. Examples of the use of figs 4.3 and 4.4 are provided in the following pages.

Note. Pressure p_1 is the initial gas pressure in the uncharged accumulator. Eqns 4.1, 4.2, 4.3 and 4.4 show that V_1 increases as p_1 decreases, hence to reduce the size of the accumulator it is desirable to have a high value of p_1. This is obtained by pre-charging the accumulator from a gas bottle to a value of p_1 that is say 10 bar below the known value of p_3. Hence p_1 is known as the pre-charge pressure.

Setting the Pressures p_2 and p_3

Fig. 4.5 shows a simple circuit using an accumulator and unloading system which permits automatic cut-out of the pump at pressure p_2 and cut-in at pressure p_3.

For an explanation of this circuit consider the pump to be pumping normally through the non-return valve B to the system and, at the same time, to be charging the accumulator C. The spring F has a lower pressure rating than K. As the system pressure rises from zero, p_3 will be reached first and F will be overcome causing spool L of the unloading differential valve to move to the left. This

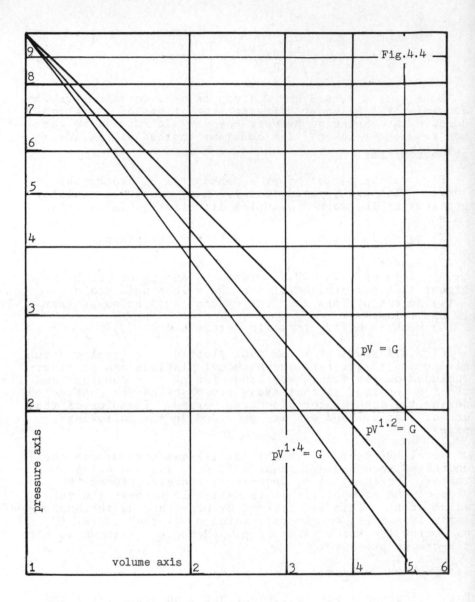

Fig.4.4

pV = G

$pV^{1.2} = G$

$pV^{1.4} = G$

pressure axis

volume axis

allows pressure p_3 (at N) to act on spool T where it is opposed by p_3 (at D) and the light spring E, hence T remains closed.

As the system pressure continues to rise pressure p_2 is reached and this pressure (at G) overcomes spring K causing poppet I to open. The resulting small oil flow through restriction J results in the pressure at A, N and P to be greater than that at G and D hence spool T moves to open and connects D to tank. T now remains open, under

the action of pressure at P and the pressure at G is reduced to near tank pressure. This results in spool H
moving .to open A to tank, via M, and so the whole pump
output goes to tank at a pressure approximately equal to
the light spring Q. The pump is now unloaded and the
system, beyond B, can only draw oil from the accumulator
C. Spring K has now reclosed poppet I.

As oil is expelled from the accumulator the system
pressure decreases until it reaches the value p_3 again.
At this point spring F closes spool L which causes the
pressure at P to be put to tank via L. Spool T now closes
due to spring E and the pressure at D rises. Hence the

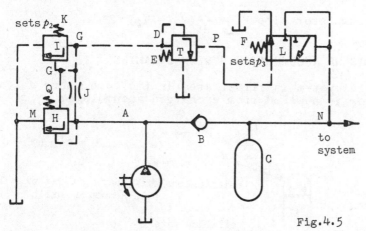

Fig.4.5

pressure at G rises to p_3 causing H to close resulting in
the pressure at A rising to (p_3 + spring Q). The pump now
delivers oil via B to the system again. This rather com-
plex arrangement permits control of the cut-in pressure
(p_3) and cut-out pressure (p_2) for the pump.

The Economics of an Accumulator

The best accumulator may be defined as the one that will
supply the maximum amount of fluid energy for a given size
accumulator. Now the maximum useful energy output is $p_3 \times$
($V_3 - V_2$) since p_3 is the pressure required to operate the
equipment and all pressures in excess of this are not useful. Such pressures are, however, necessary in order that
a gas type accumulator can operate at all. Only a dead-
weight accumulator would have a constant pressure p_3 at all
times.

Let $E = p_3 (V_3 - V_2)$

Assuming isothermal processes, $V_2 = V_3 \, p_3/p_2$, therefore

$$E = p_3 V_3 - (p_3)^2 \, \frac{V_3}{p_2}$$

119

For max.E, $dE/dp_3 = 0$, therefore

$$V_3 = V_3 \left(\frac{2p_3}{p_2}\right)$$

$$p_3 = \frac{1}{2} p_2$$

i.e. minimum pressure should be half the maximum pressure for maximum stored energy. This gives

$$E_{max.} = \frac{1}{2} p_3 V_3$$

and accumulator size $V_1 = \frac{p_2}{p_1}(V_3 - V_2)$

ANALYSIS OF ACCUMULATOR SYSTEM DYNAMICS

A simple system is illustrated in fig. 4.6 where a loaded actuator is powered from a charged accumulator

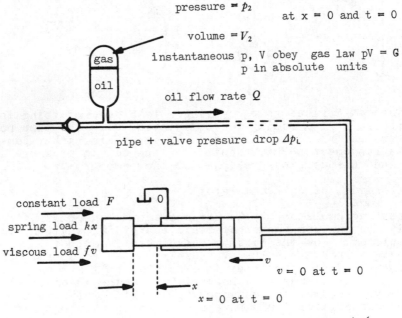

pressure $= p_2$

at x = 0 and t = 0

volume $= V_2$

instantaneous p, V obey gas law pV = G
p in absolute units

oil flow rate Q

pipe + valve pressure drop Δp_L

constant load F

spring load kx

viscous load fv

$v = 0$ at $t = 0$

$x = 0$ at $t = 0$

Fig.4.6

Note: (a) $v = dx/dt = Q/A$
(b) $Q = dV/dt$ } liquid compressibility ignored
(c) $A = dV/dx$
(d) $pV = G = $ constant

120

Consider the forces at the actuator piston

$$(p - \Delta p_L) \, A = M\frac{dv}{dt} + fv + kx + F \tag{4.5}$$

Note: F includes a term equal to (atmospheric pressure × area A).

The pipe and valve pressure losses will be assumed to be of the form $\Delta p_L = cQ^2$. Hence 4.5 may be rewritten as

$$(p - cQ^2) \, A = \frac{M}{A}\frac{dQ}{dt} + f\frac{Q}{A} + \frac{k}{A}\int Q \, dt + F$$

Now $Q = dV/dt$, $V - V_2 = \int Q \, dt$ and $pV = G$, thus

$$\frac{G}{V} - c\left(\frac{dV}{dt}\right)^2 = \frac{M}{A^2}\frac{d^2V}{dt^2} + \frac{f}{A^2}\frac{dV}{dt} + \frac{k}{A^2}(V - V_2) + \frac{F}{A}$$

which may be rewritten as

$$\frac{d^2V}{dt^2} + \frac{f}{M}\frac{dV}{dt} + \frac{cA^2}{M}\left(\frac{dV}{dt}\right)^2 + \frac{k}{M}(V - {}^{\prime}V_2) - \frac{GA^2}{M}\frac{1}{V} + \frac{FA}{M} = 0 \tag{4.6}$$

In order to obtain information about the movement of the mass M, eqn 4.6 is expressed in terms of piston displacement x.

Since the volume $V = V_2$ at $x = 0$ and $t = 0$, then

$$V = V_2 + Ax$$

Eqn 4.6 can be stated as

$$M \, D^2x + f \, Dx + cA^3(Dx)^2 + kx = \frac{GA}{V_2 + Ax} - F \tag{4.7}$$

where $D = d/dt$. The solution of this differential equation requires the use of a computer and the following values form the limits within which most calculations would be made.

M : 50 to 50×10^3 kg
f : 10^{-1} to 10^2 N s m^{-1}
c*: 6×10^{12} to 12×10^8 N s^2 m^{-6}
A : 30×10^{-4} to 3000×10^{-4} m^2
k : 10^3 to 10^6 N m^{-1}
G : 10^3 to 3×10^6 N m
V$_2$: 10^{-2} to 10^2 m^3
F : 10 to 2×10^4 N
x : 0 to 20 m
t : 0 to 100 s

*Note on c: Frictional resistance in pipes may be expressed as $0.03 \rho 1Q^2/d^5$ (approximately). If the oil density is taken as 800 kg m^{-3}, then $\Delta p_L = 24Q^21/d^5$ N/m^2, therefore

$$c = 24 \frac{1}{d^5}$$

Assuming 1 is between 5 m and 100 m, and d is between 1 cm and 10 cm then in very extreme cases c will vary between 6×10^{12} for short pipes of small bore and 12×10^8 for long pipes of large bore. Remember that 1 and d are expressed in metres, and Q in m^3/s.

Figs 4.7a, b, c, show some computer-constructed graphs of distance moved (x) against time (t) for various parameter values as indicated.

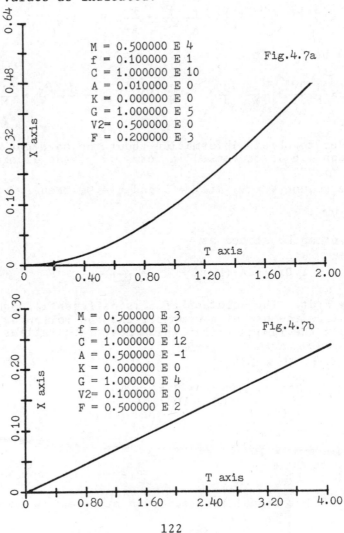

Fig.4.7a

M = 0.500000 E 4
f = 0.100000 E 1
C = 1.000000 E 10
A = 0.010000 E 0
K = 0.000000 E 0
G = 1.000000 E 5
V2= 0.500000 E 0
F = 0.200000 E 3

Fig.4.7b

M = 0.500000 E 3
f = 0.000000 E 0
C = 1.000000 E 12
A = 0.500000 E –1
K = 0.000000 E 0
G = 1.000000 E 4
V2= 0.100000 E 0
F = 0.500000 E 2

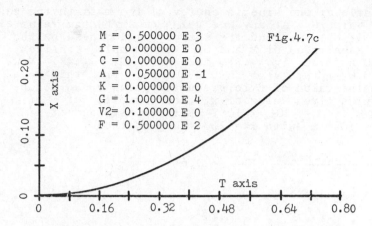

Fig.4.7c

$$M = 0.500000 \ E \ 3$$
$$f = 0.000000 \ E \ 0$$
$$C = 0.000000 \ E \ 0$$
$$A = 0.050000 \ E \ -1$$
$$K = 0.000000 \ E \ 0$$
$$G = 1.000000 \ E \ 4$$
$$V2 = 0.100000 \ E \ 0$$
$$F = 0.500000 \ E \ 2$$

Simplified Solution

For academic interest it is possible to deal with eqn 4.7 by making several assumptions that are not always justified. If viscous friction, fluid flow losses, constant resisting force and spring resistance are all zero, i.e.

$$f = c = k = F = 0$$

then eqn 4.7 reduces to

$$M \ D^2 x = \frac{GA}{V_2 + Ax}$$

Putting $D^2 x = v \ dv/dx$ gives

$$Mv \ dv = GA \frac{dx}{V_2 + Ax}$$

thus $M \frac{v^2}{2} + L = G \ \log_e (V_2 + Ax)$

where L is a constant. Now $v = 0$ and $x = 0$ at $t = 0$, therefore

$$L = G \ \log_e V_2$$

thus $v^2 = \frac{2G}{M} \log_e \left(1 + \frac{A}{V_2} x \right)$ \hfill (4.8)

Fig. 4.8 shows the relationship between $Mv^2/(2G)$ and Ax/V_2 obtained from eqn 4.8. It should be noted that Ax is the swept volume of the actuator when the piston has moved a distance x, and V_2 is the volume occupied by the gas when charged to maximum pressure by the pump. $Mv^2/(2G)$

123

is the ratio of the kinetic energy of the mass to the gas constant value G. At any one value of Ax/V_2 the value of $Mv^2/(2G)$ is obtained and the speed v will depend on the relative magnitudes of M and G. For example, consider $Ax/V_2 = 0.6$ (from fig. 4.8), then $Mv^2/(2G) = 0.48$. Now, if M = 1000 kg then $v^2 = 0.96 \times 10^{-3}$ G. From this rela- tionship the value of G to produce a given v can be ob- tained for a given value of x. For example, for $v = 10$ m s^{-1}, G = 104 × 10^3. But G = p_2V_2 and if A = 300 × 10^{-4} m^2 with x = 20 × 10^{-2} m, then

$$\frac{300 \times 10^{-4} \times 20 \times 10^{-2}}{V_2} = 0.6$$

thus $V_2 = 10 \times 10^{-3}$ m^3

and $104 \times 10^3 = p_2 \times 10 \times 10^{-3}$

$p_2 = 104 \times 10^5$ N/m^2

$= 104$ bar

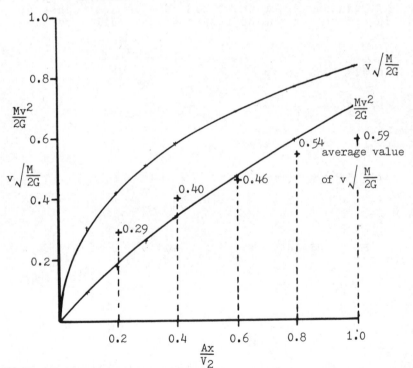

Fig.4.8

i.e. maximum pump pressure (and accumulator pressure) will need to be 104 bar. Since $p_1V_1 = p_2V_2 = G$ then a high value of G calls for a high pre-charge pressure p_1.

124

Fig. 4.8 also shows the values of $v\sqrt{(M/2G)}$ at each value of the ratio Ax/V_2. Average values of $v\sqrt{(M/2G)}$ have been indicated for values of $Ax_{max.}/V_2$ of 0.1 to 1.0. These figures will give the average piston velocity v during the stroke for different maximum stroke lengths $x_{max.}$

In accumulator problems it is usual that some information is factual (e.g. mass M and stroke $x_{max.}$) and some is obtained by reasoned assumptions, (e.g. max. p_2 based on pumps available; piston area A based on p_3 A = minimum force to drive system). An approach to problem solution would be as follows

(a) Assume p_2, p_3, A and $v_{average}$.

(b) Evaluate $v_{av.}\sqrt{(M/2G)}$ and $Ax_{max.}/V_2 (= Ax_{max.}p_2/G)$ in terms of G.

(c) By use of fig. 4.8 select a value of G that enables $v_{av.}\sqrt{(M/2G)}$ and $Ax_{max.}p_2/G$ to coincide with one of the average value points bearing in mind the fact that $Ax_{max.}/V_2$ is the ratio of maximum cylinder capacity to accumulator gas pre-charge volume, i.e. $(V_3 - V_2)/V_2$ in fig. 4.2.

(d) Now $G = p_2V_2 = p_3V_3 = p_1V_1$ where $V_3 = V_2 + Ax_{max.}$ and V_1 = accumulator size.

In this analysis it has been assumed that the gas in the accumulator obeyed the isothermal gas law $pV = G$. If, however, the gas obeys the isentropic gas law $pV^\gamma = G$ where $\gamma = 1.4$ (approx.) then eqns 4.6 and 4.7 would need to be modified and eqn 4.7 would become

$$M D^2x + f Dx + cA^3(Dx)^2 + kx = \frac{GA}{(V_2 + Ax)^\gamma} - F \quad (4.9)$$

Eqn 4.8 would become

$$v^2 = \frac{2p_2V_2}{0.4M} \left[1 - \left(1 + \frac{Ax}{V_2} \right)^{-0.4} \right] \quad (4.10)$$

THERMODYNAMIC CONSIDERATIONS

Consider the accumulator (fig. 4.9a) at pre-charge conditions represented by suffix 1. The mass of gas is given by

$$M = \frac{p_1V_1}{RT_1}$$

where R = gas constant = $C_p - C_v$

125

C_p = gas specific heat at constant pressure

C_v = gas specific heat at constant volume

The internal energy (U) of the gas is given by $U_1 = MC_vT_1$. If the gas is compressed (fig. 4.9b) then $\Delta \dot{W}$ represents the work done by the gas on the oil and ΔQ represents the heat added to the gas during the process. Both of these quantities are positive as indicated.

From the first law of thermodynamics

$$\Delta Q - \Delta W = \Delta U$$

During an accumulator charging process ΔW is negative (i.e. work done on the gas) and ΔQ is negative (i.e. heat rejected from the gas). Therefore

$$- \Delta Q_1 + \Delta W_1 = \Delta U_1$$

i.e. $\Delta W_1 = \Delta U_1 + \Delta Q_1$ $\hspace{2cm}$ (4.11)

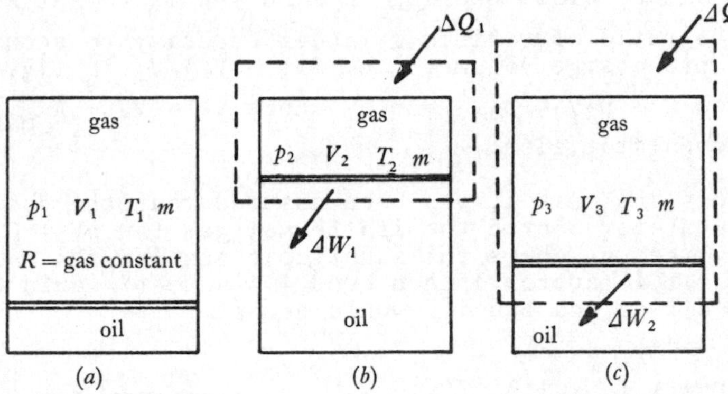

Fig.4.9

When the gas expands to expel oil from the accumulator ΔW is positive and ΔQ is positive (fig. 4.9c), therefore

$$\Delta Q_2 - \Delta W_2 = \Delta U_2$$

i.e. $\Delta W_2 = \Delta Q_2 - \Delta U_2$ $\hspace{2cm}$ (4.12)

Consider now the two extreme cases of this working process.

(1) Isothermal

$T_1 = T_2 = T_3$, so $\Delta U_1 = \Delta U_2 = 0$

From eqn 4.11

$$\Delta W_1 = \Delta Q_1 = p_1 V_1 \log_e \frac{p_2}{p_1}$$

126

From eqn 4.12

$$\Delta W_2 = \Delta Q_2 = p_1 V_1 \log_e \frac{p_2}{p_3}$$

(Note: $p_3 > p_1$.) Hence the pre-charge pressure p_1 determines the mass of gas present and the maximum pressure p_2 determines the amount of energy transfer for a given mass of gas. It should be noted that no energy is stored in the gas - it is all transferred to, and recovered from, the environment.

(2) Isentropic

$$\Delta Q_1 = \Delta Q_2 = 0$$

From eqn 4.11

$$\Delta W_1 = \Delta U_1 = m C_v (T_2 - T_1)$$

From eqn 4.12

$$\Delta W_2 = - \Delta U_2 = - m C_v (T_3 - T_2) = m C_v (T_2 - T_3)$$

In this case there is no heat transfer between the gas and the environment and all the energy is stored in the gas as increased internal energy. In practice the accumulator process is somewhere between these two extremes and some energy is stored in the gas as internal energy, and some in the environment as heat transferred.

THE USE OF AN ACCUMULATOR TO ABSORB SHOCK PRESSURES

Consider the mass of fluid in the pipe to be decelerated by the pressure created at the accumulator when valve B is closed rapidly (fig. 4.10).

$$- (\rho A L) \frac{dv}{dt} = pA$$

now $\dot{m} = - \rho \frac{dV}{dt} = \rho A v$

thus $pA = - \frac{p}{v} \frac{dV}{dt}$

and $\rho A L v\, dv = p\, dV$ (4.13)

(1) For an Isothermal Process

$pV = G$ therefore

$$\rho A L v\, dv = G \frac{dV}{V}$$

127

Integrating between initial and final conditions gives

$$\rho AL \frac{v^2}{2} = p_1 V_1 \log_e \frac{p_2}{p_3} \tag{4.14}$$

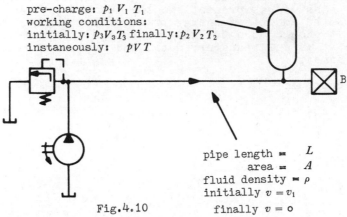

pre-charge: $p_1 \, V_1 \, T_1$
working conditions:
initially: $p_3 V_3 T_3$ finally: $p_2 \, V_2 \, T_2$
instaneously: $p \, V \, T$

pipe length = L
area = A
fluid density = ρ
initially $v = v_1$
finally $v = 0$

Fig. 4.10

(2) For an Isentropic Process

$pV^\gamma = G$ therefore

$$\rho ALv \, dv = G \frac{dV}{V^\gamma}$$

$$- \rho AL \frac{v^2}{2} = \left[\frac{GV^{1-\gamma}}{1 - \gamma} \right]_{V \, = \, V_3}^{V \, = \, V_2}$$

$$= \frac{G}{\gamma - 1} (V_3{}^{1-\gamma} - V_2{}^{1-\gamma})$$

$$= \frac{1}{\gamma - 1} (p_3 V_3 - p_2 V_2)$$

i.e. $\rho AL \dfrac{v^2}{2} = \dfrac{1}{\gamma - 1} (p_2 V_2 - p_3 V_3)$ \hfill (4.15a)

or $\quad \rho AL \dfrac{v^2}{2} = \dfrac{p_1 V_1}{\gamma - 1} \left[\left(\dfrac{p_2}{p_1} \right)^{(\gamma-1)/\gamma} - 1 \right]$ \hfill (4.15b)

if $p_3 V_3 = p_1 V_1$ i.e. isothermal process from pre-charge to working conditions.

Eqns 4.14 and 4.15b will permit V_1 (just less than accumulator volumetric capacity) to be calculated. In each case ρ, A, L. v, p_1 and p_2 need to be known.

128

It should be noted that this simple theory has neglected fluid compressibility and pipe elasticity; however, the accumulator capacity calculated will be that necessary greatly to reduce the shock pressures resulting from sudden deceleration of fluids in pipelines. The accumulator should always be fitted as close to the source of shock as possible. An empirical formula that may be used to obtain V_1 is

$$V_1 = \frac{4QLp_2}{p_2 - p_1} \times 10^{-3} \text{ litres}$$

where Q = oil flow rate l/s
      L = pipe length m
     p_2 = max. pressure bar
     p_1 = initial pressure bar

This is only suitable when the pipe diameter does not exceed 50 mm and the oil velocity at p_1 does not exceed 1.5 m/s.

WORKED EXAMPLES

1. Determine the size of the accumulator necessary to supply 5 l of oil between pressures of 200 bar and 100 bar absolute. Assume an accumulator pre-charge pressure of 90 bar.

  (a) Assuming isothermal processes

  Let V_1 be the volume of gas at pre-charge pressure p_1 = 90 bar. (fig. 4.11)
  Let V_2 be the volume of gas at maximum pressure p_2 = 200 bar. (fig. 4.11)
  Let V_3 be the volume of gas at minimum pressure p_3 = 100 bar. (fig. 4.11)

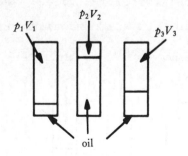

Fig.4.11

$$p_1V_1 = p_2V_2 = p_3V_3$$

thus $V_2 = \dfrac{90}{200} V_1$ and $V_3 = \dfrac{90}{100} V_1$

$$\text{Oil delivered} = V_3 - V_2 = V_1 \left(\frac{90}{100} - \frac{90}{200} \right)$$

$$= \frac{90}{200} V_1$$

therefore

$$V_1 = \frac{200}{90} \times 5 = 11.1 \text{ l}$$

(b) Assuming isentropic processes

$$p_1 V_1^{\gamma} = p_2 V_2^{\gamma} = p_3 V_3^{\gamma}$$

where $\gamma = 1.4$, therefore

$$V_2 = \left(\frac{90}{200} \right)^{1/1.4} V_1 \quad \text{and} \quad V_3 = \left(\frac{90}{100} \right)^{1/1.4} V_1$$

thus $V_3 - V_2 = 0.376 V_1$

and $V_1 = \dfrac{5}{0.376} = 11.34 \text{ l}$

Hence a 12 l capacity accumulator would be suitable.

2. In a system using a pump with a delivery of 400 ml/s, and a maximum pressure of 70 bar gauge, there is a demand for 0.8 l of oil over a period of 0.1 s at intermittent intervals. The minimum time interval between demands is 30 s. Determine the size of a suitable accumulator assuming an allowable pressure differential of 10 bar.

Without an accumulator the flow rate during demand periods would be an average of 0.8/0.1 = 8 l/s. Since the 400 ml/s pump will be delivering during a minimum period of 30 s then the accumulator stored volume cannot exceed 12 l.

With reference to fig. 4.12 and assuming isentropic processes

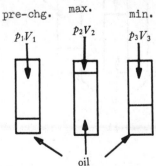

Fig.4.12

$$p_1V_1^{1.4} = p_2V_2^{1.4} = p_3V_3^{1.4}$$

although charging may be considered isothermal i.e. $p_1V_1 = p_2V_2$, therefore

$$71V_2^{1.4} = 61V_3^{1.4}$$

$$V_3 = \left(\frac{71}{61}\right)^{1/1.4} V_2 = 1.11 \ V_2$$

But $V_3 - V_2 = 0.8$ 1, therefore

$$V_3 = 1.11 \ (V_3 - 0.8)$$

$$= 8.08 \ 1$$

If the pre-charge pressure is taken as $0.9p_3$ i.e. $p_1 = 55$ bar (say) then $55V_1 = 71V_2$ (assuming isothermal charging), therefore

$$V_1 = \frac{71}{55} \ (8.08 - 0.8) = 9.4 \ 1 \text{ say } 10 \ 1 \text{ capacity}$$

The initial oil flow to the accumulator would be $V_1 - V_2$

i.e. $9.4 - 7.28 = 2.12$ 1

and this would require $2.12/0.4 = 5.3$ s.

3. Two double rod actuators work in sequence. The first unit (A) extends in 6 s and requires 50 1 of oil to complete its stroke. There is a pause of 4 s and then the second unit (B) extends in 5 s and requires 30 1 of oil. A and B retract together in 3 s. The maximum pump pressure is fixed at 200 bar, the minimum pressure for A and B to extend is 130 bar and the minimum pressure while they retract is 56 bar. The unloading valve differential is 13 bar and the interval between working cycles is 200 s. Determine the capacity of a suitable accumulator for the system and compare the pump required with that which would be necessary if no accumulator were used.

The first step is to draw a volume/time graph for the sequence to determine the oil required (fig. 4.13).

It will be assumed that the minimum pressure must not be reached before the end of the stroke of each actuator, although this has not been specified in the question. 80 1 of oil are required while the pressure falls from 200 bar to a value not below 130 bar and then a further 80 1 are required with the pressure not falling below 56 bar. If it is decided that the pressure at the end of A+ and B+ is about 140 bar (i.e. a figure greater than 130 bar), and at the end of A-B- is to be fixed by the oil volume used, then fig. 4.14 can be used to perform the calculations.

131

Starting at (1), 200 bar, and proceeding to (2), 140 bar, gives a length on the volume axis equivalent to 80 1. This length now establishes the scale on the volume axis. From (2) to (3) will involve another 80 1 on the volume axis and hence (3) is fixed.

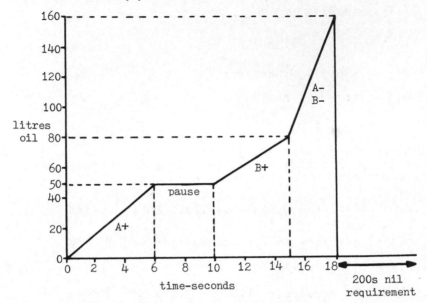

Fig.4.13

Now (2) and (3) may be moved slightly but (2) must not fall below 130 bar. The accumulator capacity is now determined by selecting a suitable pre-charge pressure at (4).

In this example set (4) at p_4 = 100 bar (pre-charge pressure). Now 80 1 scales off at $10\frac{1}{2}$ divisions i.e.1.05 units, therefore

$$V_4 = \frac{5}{1.05} \times 80 = 384 \text{ 1}$$

say a 0.4 or 0.5 m³ accumulator.

This calculation has assumed isothermal processes; had isentropic processes been assumed the volume would have been calculated using the $pV^{1.4}$ = G line (fig. 4.14).

Pump Size

(a) With accumulator
The pump will recharge the accumulator during the 200 s interval and 160 1 of oil will be required each time after the initial charge of 193 1. Therefore

132

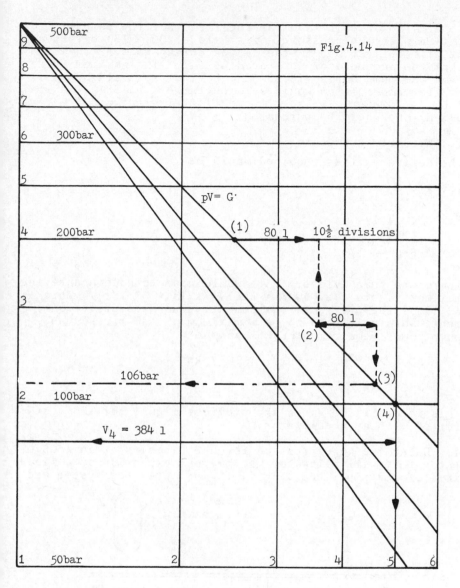

Fig.4.14

Pump delivery = $\frac{160}{200}$ = 0.8 1/s minimum

A 1 1/s pump would be suitable.

Maximum pressure = 200 bar

thus maximum power required = 200 × 10⁵ × 1 × 10⁻³ × 10⁻³ kW

$$= 20 \text{ kW}$$

133

If allowance is made for the fact that the pump will cut-in when the pressure drops below (200 - 13) bar then it will deliver oil to the system during most of the time that the accumulator is providing oil. The maximum feed from the pump during 18 s would be 18 1 so a slightly smaller accumulator would be acceptable i.e. 384 - 18 = 366 1. However, it is still most likely that a 0.4 m³ or 0.5 m³ one would be purchased.

(b) Without accumulator

The delivery rates required would be

for A+, $\frac{50}{6}$ 1/s

for B+, $\frac{30}{5}$ 1/s

for A-B-, $\frac{80}{3}$ 1/s

Therefore pump delivery rate would need to accommodate the maximum figure i.e. 80/3 = 26.7 1/s (say 30 1/s). This is 30 times the previous requirement. The pressure requirement would only need to be a maximum of 130 bar so the pump power requirement would now be

$$30 \times 10^{-3} \times 130 \times 10^5 \times 10^{-3} \text{ kW}$$

$$= 390 \text{ kW}$$

This is nearly 20 times the previous figure for the system fitted with an accumulator.

4. Determine an expression for the capacity of an accumulator that will allow for the thermal expansion of a fluid in a closed pipeline. Fig. 4.15 indicates the system and

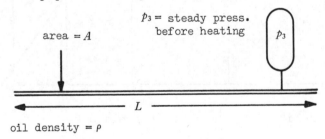

area = A

p_3 = steady press. before heating

p_3

L

oil density = ρ

Fig.4.15

p_2 is to be the maximum pressure resulting from the expansion. Neglect the expansion of the pipe itself and assume a pre-charge pressure of p_1 for the accumulator. p_3 = steady pressure before heating.

Let accumulator volume (gas) be V_3 at p_3
Let accumulator volume (gas) be V_2 at p_2 (max. pressure)

134

Let oil coefficient of thermal expansion be β (volumetric)
Let oil temperature rise be Δt

Now $V_2 = V_3 - AL\beta\Delta t$

and $V_2 = V_3 \left(\dfrac{p_3}{p_2}\right)^{1/\gamma}$ for isentropic process

therefore

$$V_3 = \frac{AL\beta\Delta t}{1 - (p_3/p_2)^{1/\gamma}}$$

Now $p_1 V_1{}^\gamma = p_3 V_3{}^\gamma$ for isentropic process

therefore

$$V_1 = \frac{AL\beta\Delta t}{1 - (p_3/p_2)^{1/\gamma}} \times \left(\frac{p_3}{p_1}\right)^{1/\gamma} = \begin{array}{l}\text{accumulator}\\ \text{capacity}\end{array}$$

or for isothermal processes

$$V_1 = \frac{AL\beta\Delta t}{p_1\left(\dfrac{1}{p_3} - \dfrac{1}{p_2}\right)}$$

5. Fig. 4.16 shows a pump discharging through a pipeline
to tank. An accumulator is connected close to the pump
in order to damp the pulsations due to the delivery
'ripple' from the pump.

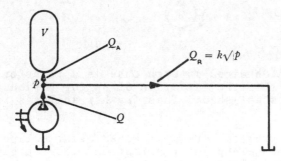

Fig.4.16

  Determine an expression relating the pump delivery pressure fluctuation (δp) to the flow ripple (δQ)

  (a) when all the inertia effects of the oil are neglected.

(b) when the mass of oil in the pipeline to tank is accounted for.

(c) when a laminar resistance is introduced into the accumulator line (fig. 4.17) and all inertia effects are neglected.

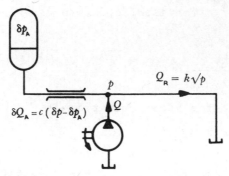

Fig.4.17

(d) when a laminar resistance is introduced into the accumulator line (fig. 4.18) and the mass of oil in the accumulator and its pipeline is accounted for.

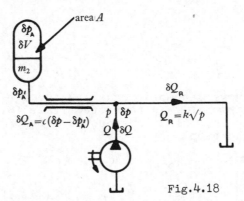

Fig.4.18

The expression obtained in each case is a transfer function and from this the variation of δp for various forms of δQ may be established. This further analysis is dealt with later.

Note: the problems are approached using a linearisation technique whereby small variations about a given condition are considered.

If P is a function of variable Q, R, S and T then

$$P = \phi(Q, R, S, T)$$

and $\quad \delta P = \dfrac{\partial P}{\partial Q}\, \delta Q + \dfrac{\partial P}{\partial R}\, \delta R + \dfrac{\partial P}{\partial S}\, \delta S + \dfrac{\partial P}{\partial T}\, \delta T$

136

(a) Consider the system indicated in fig. 4.16 with the following assumptions.

(1) The resistance to flow may be expressed by $Q_R = k\sqrt{p}$.

(2) The mean pump delivery is Q at pressure p and a small variation of Q takes place causing a pressure variation δp. The relationship $\delta p/\delta Q$ is required.

(3) With no variation in flow $Q = Q_R$ and $Q_A = 0$.

(4) The accumulator volume (gas) is V for condition (3).

(5) The oil mass may be neglected.

(6) The gas law for the accumulator is $pV^\gamma = G = $ constant.

Now $Q + \delta Q = Q_R + \delta Q_R + \delta Q_A$

thus $\delta Q = \delta Q_R + \delta Q_A$ $\qquad\qquad\qquad\qquad\qquad$ (4.16)

Also $\delta p = \dfrac{\partial p}{\partial Q} \delta Q_R$

since $Q = Q_R$ and $Q_A = 0$

and $\dfrac{\partial Q}{\partial p} = \dfrac{1}{2} \dfrac{k}{\sqrt{p}} = \dfrac{Q}{2p}$

since $Q = Q_R$. Therefore

$$\delta p = \frac{2p}{Q} \delta Q_R \qquad\qquad\qquad\qquad (4.17)$$

Also $\delta V = \dfrac{\partial V}{\partial p} \delta p$

and $\dfrac{\partial V}{\partial p} = -\dfrac{V}{\gamma p}$ from $pV^\gamma = G$

therefore

$$\delta V = -\frac{V}{\gamma p} \delta p \qquad\qquad\qquad\qquad (4.18)$$

but $\delta Q_A = -\dfrac{d}{dt} (V + \delta V)$

thus $\delta Q_A = -\dfrac{d}{dt} (\delta V)$ $\qquad\qquad\qquad\qquad$ (4.19)

Eqns 4.18 and 4.19 give

$$\delta Q_A = \frac{V}{\gamma p} \frac{d}{dt} (\delta p) \qquad\qquad\qquad\qquad (4.20)$$

Eqns 4.16, 4.17 and 4.20 give

$$\delta Q = \frac{Q}{2p} \delta p + \frac{V}{\gamma p} \frac{d}{dt} (\delta p)$$

since the initial condition for incremental values is zero, then (using Laplace 's')

$$\frac{\delta p}{\delta Q} = \frac{1}{\dfrac{V}{\gamma p} s + \dfrac{Q}{2p}} = \frac{\gamma p}{V} \left(\frac{1}{s + \dfrac{\gamma Q}{2V}} \right) \qquad (4.21a)$$

Putting $\gamma Q/(2V) = \omega_1$ makes eqn 4.21a into

$$\frac{\delta p}{\delta Q} = \frac{\gamma p}{V} \left(\frac{1}{s + \omega_1} \right) \qquad (4.21b)$$

or $\qquad \dfrac{\delta p}{\delta Q} = \dfrac{2p}{Q} \left(\dfrac{1}{\dfrac{1}{\omega_1} s + 1} \right)$

Note: Had the pipe resistance been laminar (i.e. $Q_R = kp$) then eqn 4.21b would be the same except that $\omega_1 = \gamma Q/V$.

(b) If, now, the mass of oil in the pipeline to tank is considered as M_R but the mass associated with the accumulator is neglected   then

$$(p + \delta p) \, a = M_R \frac{dv}{dt} + \left(\frac{Q_R + \delta Q_R}{k} \right)^2 a$$

where a = pipe section area
$\qquad v$ = mean pipe oil speed

Therefore

$$\delta p = \frac{M_R}{a} \frac{dv}{dt} + \frac{2 Q_R \, \delta Q_R}{k^2}$$

assuming δQ_R^2 can be neglected.

Now $\quad v = \dfrac{Q_R + \delta Q_R}{a}$

therefore

$$\frac{dv}{dt} = \frac{1}{a} \frac{d}{dt} (\delta Q_R)$$

so $\qquad \delta p = \dfrac{M_R}{a^2} \dfrac{d}{dt} (\delta Q_R) + \dfrac{2 Q_R}{k^2} \delta Q_R$

but $Q_R = Q$ and $k^2 = Q_R^2/p$, hence, for zero initial conditions

138

$$\delta p_. = \left[\frac{M_R}{a^2} s + \frac{2p}{Q}\right] \delta Q_R \qquad (4.22)$$

Eqn 4.20 still holds, i.e.

$$\delta Q_A = \frac{V}{\gamma p} \frac{d}{dt} (\delta p)$$

hence, eqns 4.16, 4.20 and 4.22 give

$$\delta Q = \frac{\delta p}{\dfrac{M_R}{a^2} s + \dfrac{2p}{Q}} + \frac{V}{\gamma p} s \, \delta p$$

i.e.
$$\delta Q = \left[\frac{1 + \dfrac{M_R}{a^2} \dfrac{V}{\gamma p} s^2 + \dfrac{2V}{\gamma Q} s}{\dfrac{M_R}{a^2} s + \dfrac{2p}{Q}}\right] \delta p$$

Put $(\gamma p/V)(a^2/M_R) = \omega_2^2$ with $Q/(2V) = \omega_1$, then

$$\frac{\delta p}{\delta Q} = \frac{\dfrac{2p}{Q}\left[\dfrac{\omega_1}{\omega_2^2} s + 1\right]}{\left[\dfrac{1}{\omega_2^2} s^2 + \dfrac{1}{\omega_1} s + 1\right]} \qquad (4.23)$$

Note: Had the pipe resistance been laminar (i.e. $Q_R = kp$) then eqn 4.23 would have been the same except that $\omega_1 = \gamma Q/V$.

(c) Consider a case in which a laminar resistance ($\delta Q_A = c(\delta p - \delta p_A)$) exists in the accumulator line. δp_A is variation of gas pressure in accumulator associated with δp at pump (fig. 4.17). Neglecting all mass inertia effects of the oil in the system

$$\delta Q_A = c(\delta p - \delta p_A)$$

and from eqn 4.20

$$\delta Q_A = \frac{V}{\gamma p} \frac{d}{dt} (\delta p_a)$$

$$= \frac{V}{\gamma p} \frac{d}{dt} \left[\delta p - \frac{\delta Q_A}{c}\right]$$

139

with zero initial conditions

$$\left(\frac{\gamma p}{V} + \frac{1}{c}\, s\right)\, \delta Q_A = s\, \delta p \tag{4.24}$$

and from eqn 4.17 $\delta p = (2p/Q)\delta Q_R$ where pipe mass of fluid is neglected, therefore

$$\delta Q = \left[\frac{s}{\frac{\gamma p}{V} + \frac{1}{c}\, s} + \frac{Q}{2p}\right]\, \delta p$$

$$\frac{\delta p}{\delta Q} = \frac{\frac{\gamma p}{V} + \frac{1}{c}\, s}{s\left[1 + \frac{Q}{2pc}\right] + \frac{\gamma Q}{2V}} = \frac{\frac{2p}{Q}\left[\frac{\gamma Q}{2V} + \frac{Q}{2pc}\, s\right]}{s\left[1 + \frac{Q}{2pc}\right] + \frac{\gamma Q}{2V}}$$

$$= \frac{2p}{Q}\, \frac{\left[\omega_1 + \frac{Q}{2pc}\, s\right]}{\omega_1 + \left[1 + \frac{Q}{2pc}\right]\, s}$$

Note: $c = a^2/(8\pi\mu L_1)$, where a = pipe sectional area, μ = oil viscosity (dynamic), L_1 = pipe length (pump to accumulator).

(d) Consider this case where M_1 is the mass of oil in the pipeline to the accumulator and M_2 is the mass of oil in the accumulator. Let the oil surface area in the accumulator be A and assume this to be a constant. Normal conditions are p and Q at the pump, p_A and V in the accumulator, Q_R in the pipeline to tank. Variations are: δp, δQ, δp_A, δV, δQ_R with δQ_A to the accumulator $\delta p_A'$ is the pressure fluctuation at the line of separation between mass M_1 and mass M_2 (see fig. 4.18).

In the accumulator and line

$$\left[(p + \delta p) - (p + \delta p_A')\right]a = M_1 \frac{dv}{dt} + \frac{\delta Q_A}{c}\, a$$

thus $\delta p - \delta p_A' = \dfrac{M_1}{a^2} \dfrac{d}{dt}\, (\delta Q_A) + \dfrac{1}{c}\, \delta Q_A$

and $(\delta p_A' - \delta p_A)\, A = \dfrac{M_2}{A} \dfrac{d}{dt}\, (\delta Q_A)$

and $(\delta p_A' - \delta p_A)\, A = \dfrac{M_2}{A} \dfrac{d}{dt}\, (\delta Q_A)$

thus $\delta p - \delta p_A = \left(M_1 \dfrac{A^2}{a^2} + M_2 \right) \dfrac{1}{A^2} \dfrac{d}{dt} (\delta Q_A) + \dfrac{1}{c} (\delta Q_A)$

From eqn 4.20

$$\delta Q_A = \dfrac{V}{\gamma p} \dfrac{d}{dt} (\delta p_A)$$

thus $\delta p - \dfrac{\gamma p}{Vs} \delta Q_A = \left[\dfrac{M_A}{A^2} s + \dfrac{1}{c} \right] \delta Q_A$

where $M_A = [M_1 (A^2/a^2) + M_2] =$ equivalent oil mass in accumulator and $s =$ Laplace operator. Therefore

$$s \, \delta p = \left[\dfrac{M_A}{A^2} s^2 + \dfrac{1}{c} s + \dfrac{\gamma p}{V} \right] \delta Q_A \qquad (4.25)$$

Also $\delta p = \dfrac{2p}{Q} \delta Q_R$ from eqn 4.17

therefore

$$\delta Q = \left[\dfrac{Q}{2p} + \dfrac{s}{\dfrac{M_A}{A^2} s^2 + \dfrac{1}{c} s + \dfrac{\gamma p}{V}} \right] \delta p$$

$$= \dfrac{\dfrac{Q}{2p} \dfrac{M_A}{A^2} s^2 + \left(1 + \dfrac{Q}{2pc} \right) s + \dfrac{Q\gamma}{2V}}{\dfrac{M_A}{A^2} s^2 + \dfrac{1}{c} s + \dfrac{\gamma p}{V}} \, \delta p$$

therefore

$$\dfrac{\delta p}{\delta Q} = \dfrac{2p}{Q} \dfrac{s^2 + \dfrac{\omega_3{}^2}{\omega_1} \dfrac{Q}{2pc} s + \omega_3{}^2}{s^2 + \dfrac{\omega_3{}^2}{\omega_1} \left(1 + \dfrac{Q}{2pc} \right) + \omega_3{}^2}$$

where $\omega_3{}^2 = \gamma p A^2 / (V M_A)$. If

$$\zeta = \dfrac{1}{2} \dfrac{\omega_3}{\omega_1} \dfrac{Q}{2pc}$$

then $\dfrac{\delta p}{\delta Q} = \dfrac{2p}{Q} \dfrac{s^2 + 2\zeta\omega_3 s + \omega_3{}^2}{s^2 + \left(\dfrac{\omega_3{}^2}{\omega_1} + 2\zeta\omega_3 \right) s + \omega_3{}^2}$

141

FURTHER EXAMPLES

1. A hydraulic press acts over a distance of 120 mm at
a rate of 60 mm/s producing a force of 150 kN. The pump
supplying oil to drive the press has a relief valve set
at 200 bar. Determine the size of the accumulator that
would be required for use with the press if the system
pressure must not be allowed to fall below 130 bar.
Assume a pre-charge pressure of 100 bar with isothermal
charging and isentropic discharging.

   What capacity pump would be required for this system
if no accumulator were to be included?

                            (7.7 1, say 8 1; 0.7 1/s)

2. A system is required to deliver 500 ml of oil over a
period of 0.1 s with a minimum time between demands of
30 s. Determine the necessary accumulator size if the
pump associated with the system has a delivery of 200 ml/s
and a maximum delivery pressure of 75 bar. The minimum
system pressure is to be 10 bar.

   What capacity pump would be required for this system if
no accumulator were to be included?

                        (1.5 1 with p_1 = 8 bar; 5 1/s)

3. Fig. 4.19 shows the volume of oil and minimum pressure
required by a system during its working cycle which starts
once every 4 minutes. Determine the size of a suitable
accumulator for the system assuming a maximum pump pressure of 300 bar and selecting a pre-charge pressure to
suit.

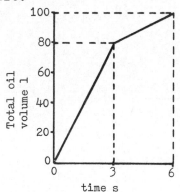 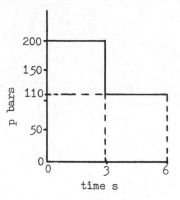

Fig.4.19

   Compare the power requirement of the pump used for this
system with that which would be required if no accumulator
was fitted (use a graphical solution).

        (350 1 (with p_1 = 150 bar); 12.8 kW c.f. 533 kW)

4. In a particular case the charging process for an accumulator may be assumed to be isothermal and the discharge process isentropic. Show that, for maximum useful energy output, $p_3/p_2 = 0.466$ and the accumulator size is given by

$$V_1 = \gamma \frac{p_2}{p_1} (V_3 - V_2)$$

where p_1 = pre-charge pressure associated with volume V_1
           of gas
      p_2 = maximum charge pressure associated with V_2
      p_3 = minimum working pressure associated with V_3
      γ = gas index = 1.4 in this case

5. A fully charged accumulator, with a gas volume V_1 at maximum pressure p_1, supplies oil to a positive-displacement, fixed capacity, hydraulic motor. The motor has a pure inertia load J and all losses can be neglected. Show that the motor shaft speed ω is related to the accumulator gas pressure p at any instant by the following expressions.

(a) $\omega^2 = \dfrac{2p_1V_1}{J} \log_e \dfrac{p_1}{p}$ for an isothermal gas expansion

(b) $\omega^2 = \dfrac{2p_1V_1}{J(\gamma - 1)} \left[1 - \left(\dfrac{p}{p_1}\right)^{\gamma-1/\gamma} \right]$ for an isentropic gas expansion

6. The simple circuit shown in fig. 4.20 consists of a fixed displacement pump, an accumulator, a length of pipe finishing with a fixed area control valve. The flow in

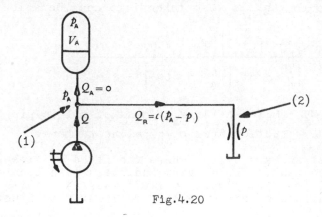

Fig.4.20

the pipe is laminar and values shown are for mean conditions when the pressure in the accumulator is p_A associated with a gas volume V_A. The pressure immediately upstream of the valve is p and the discharge is to tank. Show that small fluctuations of pump delivery pressure

143

(δp_A) due to delivery rate fluctuations (δQ) may be expressed by the transfer function

$$\frac{\delta p_A}{\delta Q} = \frac{\delta p_A}{V_A} \frac{1}{s + \omega_1 \left(\dfrac{2p_A}{p_A + p}\right)}$$

where $\omega_1 = \gamma Q_R / (2V_A)$, γ = gas law index in $p_A V_A{}^\gamma$ = constant and Q_R = flow through the control valve for pressure drop p across it (i.e. $Q_R = k\sqrt{p}$). (Neglect the mass of the oil in the system).

7. If question 8 is reworked, making allowance for the mass of oil in the pipeline to the valve (points 1 to 2) show that

$$\frac{\delta p_A}{\delta Q} = \frac{\delta p_A}{V_A} \frac{s + \dfrac{1}{\alpha} \dfrac{\omega_2{}^2}{\omega_1}}{s^2 + \dfrac{1}{\alpha} \dfrac{\omega_2{}^2}{\omega_1} s + \omega_2{}^2}$$

where $\omega_2{}^2 = \gamma p_A a^2 / (V_A m)$, $\alpha = 2p_A / (p_A + p)$, a = pipe section area and m = mass of oil in pipe between pump and valve (1 to 2).

8. Reconsider worked example 5d. If the mass of oil in the pipeline to tank is also taken into consideration, show that

$$\frac{\delta p}{\delta Q} = \frac{2p}{Q} \frac{\left[\dfrac{\omega_2{}^2}{\omega_1} s + 1\right]\left[s^2 + 2\zeta\omega_3 s + \omega_3{}^2\right]}{\left[\dfrac{M_E}{M_A} s^2 + \left(\dfrac{\omega_3{}^2}{\omega_1} + 2\zeta\omega_3\right) s + \omega_3{}^2\right]}$$

where $M_E = [M_A + M_R (A^2/a^2)]$, M_R = mass of oil in pipeline to tank and a = section area of pipeline to tank.

9. Fig. 4.21 shows a system where the fixed-orifice valve discharges to tank. The diagram indicates mean conditions for a flow Q from the pump. If fluid mass is ignored show that fluctuations of pressure δp_2 are related to fluctuations δp_1 by

(a) $\dfrac{\delta p_2}{\delta p_1} = \dfrac{p_2}{p_1} \dfrac{\omega_1}{s + \omega_1}$ for laminar pipe flow

(b) $\dfrac{\delta p_2}{\delta p_1} = \dfrac{2p_2}{p_1 + p_2} \dfrac{\omega_1}{s + \omega_1}$ for turbulent pipe flow

144

where $\omega_1 = \gamma Q/(2V)$, flow $Q_A = 0$ and $Q_R = k_R\sqrt{p_2}$.

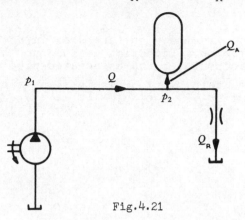

Fig.4.21

10. If, in question 9, the mass of fluid in the pipe from the pump to the accumulator and in the accumulator is considered, show that, for turbulent pipe flow and a laminar restrictor is the accumulator pipe (fig. 4.22)

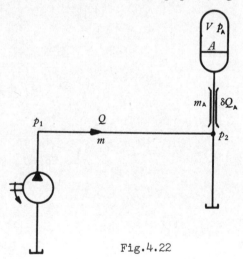

Fig.4.22

$$\frac{\delta p_2}{\delta p_1} = \frac{p_2}{p_1} \cdot \frac{(s^2 + 2\zeta\omega_3 s + \omega_3{}^2)}{\left[\dfrac{\omega_1}{\omega_2{}^2} s + 1\right]\left[s^2 + \left(\dfrac{\omega_3{}^2}{\omega_1} + 2\zeta\omega_3\right) s + \omega_3{}^2\right]}$$

where $\omega_1 = \dfrac{\gamma Q}{2V}$ $\omega_2{}^2 = \dfrac{\gamma p_1 a^2}{Vm}$

145

$$\omega_3{}^2 = \frac{\gamma p_2 A^2}{V m_A} \qquad \zeta = \frac{1}{2} \frac{\omega_3}{\omega_1} \frac{Q}{2 p_2 c}$$

a = pipe section area, A = surface area in accumulator, m = oil mass in pipe up to junction with accumulator, m_A = oil mass in accumulator and associated pipe (equivalent mass) and c = coefficient of laminar restrictor in accumulator pipe, i.e. $\delta Q_A = c(\delta p_2 - \delta p_A)$.

5   BLOCK DIAGRAMS AND SIGNAL FLOW DIAGRAMS

These are two methods of representing a dynamic system and both help with the understanding of a system in detail. Only linear systems can be represented in these diagrams (unless a non-linear system is linearised first) and from the diagrams the system transfer function can be readily obtained. This approach then leads to a study of system stability and steady-state frequency response. Before either of the diagrams can be constructed it is necessary to obtain the equations which represent the system, but it is not necessary to combine these to give a single equation. A series of examples is given which illustrate the construction and use of these diagrams.

WORKED EXAMPLES

1. An illustrative example, which is familiar to most readers, is given below (fig. 5.1). A spring/mass/damper system has an input displacement x and a resulting output y. The block diagram for the system is to be constructed and the transfer function obtained.

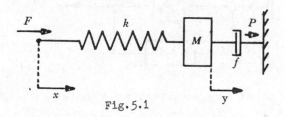

Fig.5.1

The force F, applied to the system is given by

F = k (x - y)

This force accelerates the mass and provides the force P reacting on the wall. Therefore

P = F - M $\overset{''}{y}$

and P = f $\overset{'}{y}$

Working in the s domain gives

F(s) = k [X(s) - Y(s)]

P(s) = F(s) - Ms²Y(s)

P(s) = f s Y(s)

This gives the block diagram shown in fig. 5.2.

The output from a block is equal to the product of the input and the term inside the block. Summing points and take-off points are self-evident. Since the two summing

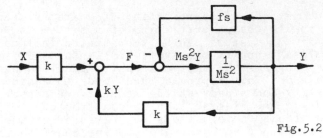

Fig.5.2

points are adjacent, and no block (except unity) appears between them, then they can be combined, making the appearance of F unnecessary. Forward paths have arrows to the right, in this example, and feedback paths have arrows to the left. The two parallel feedback paths may be combined to give a single one with a transfer function

$\boxed{k + f s}$. This gives the block diagram of fig. 5.3.

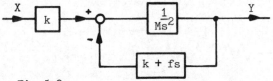

Fig.5.3

The summing point may be moved to the left of the $\boxed{k}$ block to give fig. 5.4.

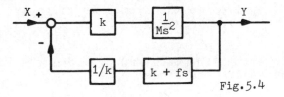

Fig.5.4

Adjacent blocks may be combined to give fig. 5.5. This final block diagram will give the transfer function of the system. Referring to fig. 5.6

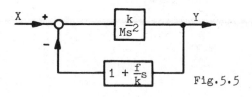

Fig.5.5

148

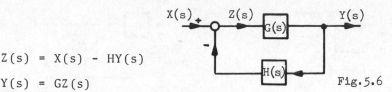

$$Z(s) = X(s) - HY(s)$$

$$Y(s) = GZ(s)$$

Fig.5.6

therefore

$$\frac{Y}{G}(s) = X(s) - HY(s)$$

$$\frac{Y}{X}(s) = \frac{G(s)}{1 + GH(s)}$$

where G(s) = forward path transfer function
       H(s) = feedback path transfer function

Hence it is shown that

$$\frac{Y}{X}(s) = \frac{\dfrac{k}{Ms^2}}{1 + \dfrac{k}{Ms^2}\left(1 + \dfrac{f}{k}\,s\right)}$$

$$\frac{Y}{X}(s) = \frac{k}{Ms^2 + f\,s + k}$$

This simple example has been worked in great detail to illustrate some of the points about block diagrams illustrated in fig. 5.7.

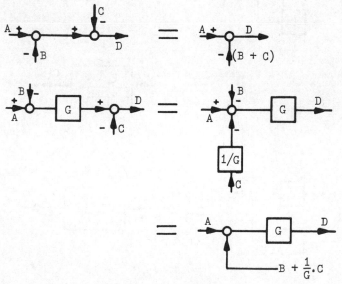

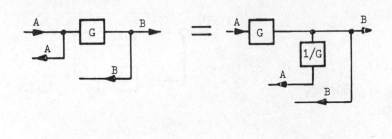

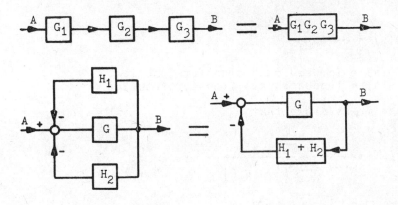

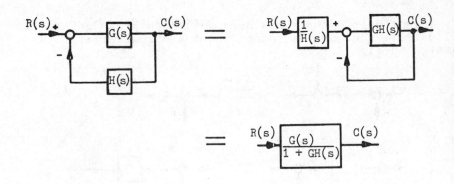

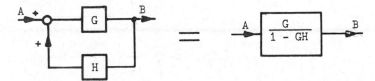

Fig.5.7

2. Reduce the block diagram of fig. 5.8 to a form containing one forward path and one feedback path, and determine the form of the closed loop transfer function relating the output swept volume rate of a motor [Y(s)] to the variable input swept volume rate of a pump [X(s)].

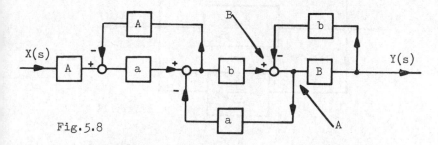

Fig. 5.8

Step 1: Move take-off point A to the right and summing point B to the left, giving fig. 5.9.

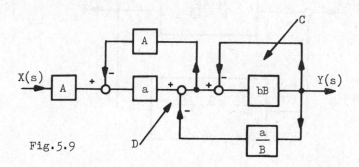

Fig. 5.9

Step 2: Remove the minor loop and move summing point D to left, giving fig. 5.10.

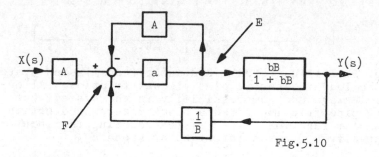

Fig. 5.10

Step 3: Move take-off point E to right and summing point F to left, giving fig. 5.11.

151

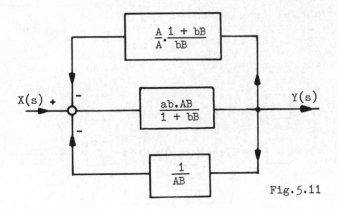

Fig.5.11

Step 4: Combine the feedback loops to give fig. 5.12, which will give the transfer function

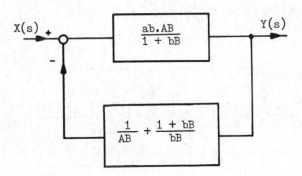

Fig.5.12

$$\frac{Y(s)}{X(s)} = \frac{ab\ AB(s)}{1 + ab + b\ B(s) + a\ A(s)\ (1 + b\ B(s))}$$

$$= \left(1 + \frac{1}{a\ A(s)} + \frac{1}{b\ B(s)} + \frac{1}{AB(s)} + \frac{1}{ab\ AB(s)}\right)^{-1}$$

3. Fig. 5.13 illustrates a hydraulic transmission system where the load is inertia plus viscous friction. Allowing for leakage and compressibility in the system, but ignoring pipe friction, draw a block diagram and obtain the transfer function of the system relating the input pump capacity C_p to the output motor speed ω_m.

Pump pressure rise Δp = motor pressure drop Δp

Pump leakage = $\lambda_p\ \Delta p$

152

Compressed volume $= \dfrac{V_O}{B} \Delta p$

on high pressure side of system, therefore

$$Q = \omega_p C_p - \lambda_p \, \Delta p - \dfrac{V_O}{B} \, \dot{\Delta p}$$

and $\quad \omega_m C_m = Q - \lambda_m \, \Delta p$

For the load

$$\eta_m \omega_m C_m \Delta p = T_m \omega_m$$

where η_m = motor mechanical efficiency

$\quad T_m$ = motor torque

But $\quad T_m = J \, \dot{\omega}_m + f \, \omega_m$

fluid volume V_0
Bulk Modulus B

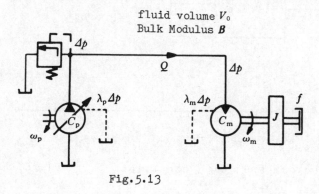

Fig.5.13

Expressing these equations in the s domain gives

$$Q = \omega_p C_p - \left[\lambda_p + \dfrac{V_O}{B} \, s \right] \Delta p$$

$$\Delta p = \dfrac{Q}{\lambda_m} - \dfrac{\omega_m C_m}{\lambda_m}$$

$$\eta_m \Delta p C_m = (Js + f) \, \omega_m$$

$$\omega_m C_m = \dfrac{\eta_m C_m^{\,2}}{Js + f} \, \Delta p$$

This gives a block diagram shown in fig. 5.14 which can be
rearranged to give fig. 5.15. This gives the transfer
function

153

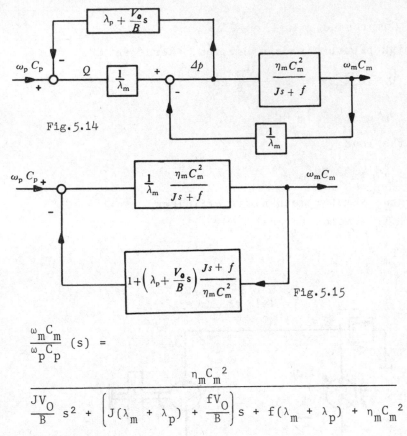

Fig.5.14

Fig.5.15

$$\frac{\omega_m C_m}{\omega_p C_p}(s) =$$

$$\frac{\eta_m C_m{}^2}{\dfrac{JV_0}{B}s^2 + \left[J(\lambda_m + \lambda_p) + \dfrac{fV_0}{B}\right]s + f(\lambda_m + \lambda_p) + \eta_m C_m{}^2}$$

which should be compared with eqn 2.16.

SIGNAL FLOW DIAGRAMS

Such diagrams are alternatives to block diagrams and are frequently easier to draw from the system equations, and, with the aid of Mason's formula, the system transfer function can be obtained quickly, even for very compli- cated systems. A signal flow diagram consists of 'nodes' which represent the system variables, and 'directed line segments' which bear the relationship between the vari- ables (fig. 5.16). Here

Fig.5.16

$$B(s) = A(s)G_1(s) - C(s)H(s)$$

154

$$C(s) = B(s)G_2(s)$$

WORKED EXAMPLES

4.   Reconsider example 1 where the equations were

$$F(s) = k[X(s) - Y(s)]$$

$$P(s) = F(s) - Ms^2Y(s)$$

$$P(s) = f s Y(s)$$

The variables are $X(s)$, $F(s)$, $P(s)$ and $Y(s)$ and the signal flow diagram is as shown in fig. 5.17.

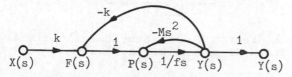

Fig.5.17

i.e. $F(s) = kX(s) - kY(s)$

$P(s) = F(s) - Ms^2Y(s)$

$Y(s) = \dfrac{1}{fs} P(s)$

The signal flow diagram can obviously be obtained directly from the block diagram when the latter is available, but no attempt should be made to construct this first in order to obtain the signal flow diagram. The latter is always obtainable directly from the system equations. It should be noted that the given signal flow diagram is not the only possible one for this system but it does include all the variables.

The system equations could be rearranged as
$$F(s) = k(X(s) - Y(s))$$

$$fsY(s) = F(s) - Ms^2Y(s)$$

i.e. $Y(s) = \dfrac{F(s)}{Ms^2 + fs}$

giving fig. 5.18.

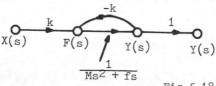

Fig.5.18

5. This involves constructing a signal flow diagram from the block diagram given in example 2. The block diagram is reproduced in fig. 5.19 and additional variables have been inserted for identification purposes. The (s), indicating s-domain working, has been omitted for clarity.

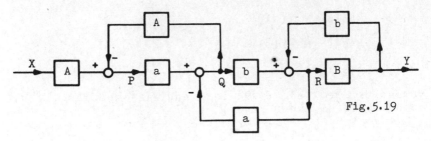

Fig.5.19

From the block diagram it is seen that the signal flow diagram can be drawn as shown in fig. 5.20.

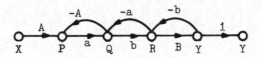

Fig.5.20

6. This employs the same transmission system as specified in example 3. The object is to obtain the signal flow diagram from the system equations, which are as follows.

$$Q = \omega_p C_p - \lambda_p \Delta p - \frac{V_0}{B} \Delta \dot{p}$$

$$\omega_m C_m = Q - \lambda_m \Delta p$$

$$T_m \omega_m = \omega_m C_m \Delta p$$

$$T_m = J \dot{\omega}_m + f \omega_m$$

In Laplace form

$$Q(s) = \omega_p C_p(s) - \left[\lambda_p + \frac{V_0}{B} s \right] \Delta p(s)$$

$$\Delta p(s) = \frac{Q(s)}{\lambda_m} - \frac{\omega_m C_m}{\lambda_m} (s)$$

$$\omega_m C_m(s) = \frac{\eta_m C_m^2}{Js + f} \Delta p(s)$$

Giving the signal flow diagram of fig. 5.21.

156

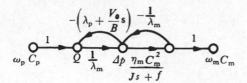

Fig. 5.21

Obtaining the System Transfer Function from the Signal Flow Diagram

For relatively simple cases this can be done by a series of steps which reduce the diagram to a single loop diagram from which the transfer function is quickly obtained. For more complicated systems, involving many loops, the application of the Mason formula is required.

Simple Case - Example 7

Obtain the transfer function for a system represented by the signal flow diagram in fig. 5.22.

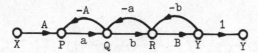

Fig. 5.22

Step 1:   Eliminate the right-hand loop by

$$R = bQ - bY \text{ and } Y = BR$$

therefore

$$Y = \frac{b}{\frac{1}{B} + b} Q$$

and $Q = aP - aR = aP - \dfrac{a}{B} Y$

giving fig. 5.23.

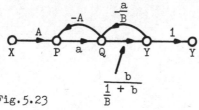

Fig. 5.23

Step 2:   Eliminate the right-hand loop by

$$Q = aP - \frac{a}{B} Y \text{ and } Y = \frac{b}{\frac{1}{B} + b} Q$$

157

therefore

$$Y\left(\frac{a}{B} + \frac{\frac{1}{B} + b}{b}\right) = aP$$

thus $Y = \dfrac{a}{\dfrac{a}{B} + \dfrac{1}{Bb} + 1} P$

and $P = AX - AQ$

therefore

$$P = AX - A\left(\frac{\frac{1}{B} + b}{b}\right)Y = AX - \left(\frac{A}{bB} + A\right)Y$$

giving fig. 5.24.

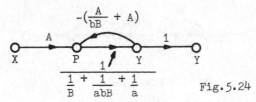

Fig.5.24

This gives

$$P = AX - \left(\frac{A}{bB} + A\right)Y$$

and $Y = \dfrac{1}{\dfrac{1}{B} + \dfrac{1}{abB} + \dfrac{1}{a}} P$

therefore

$$\left(\frac{1}{B} + \frac{1}{abB} + \frac{1}{a} + \frac{A}{bB} + A\right)Y = AX$$

and $\dfrac{Y}{X}(s) = \dfrac{1}{1 + \dfrac{1}{aA} + \dfrac{1}{bB} + \dfrac{1}{AB} + \dfrac{1}{abAB}}$

Mason's Formula

This may be stated as

$$\frac{Y}{X}(s) = \frac{M_1\Delta_1 + M_2\Delta_2 + \dots}{\Delta}$$

where M_1 = gain of forward path 1

158

M_2 = gain of forward path 2
    etc.

    Δ = 1 - $\sum$(all loop gains) + $\sum$(gain products of all combinations of 2 non-touching loops) - $\sum$(gain products of all combinations of 3 non-touching loops) + $\sum$ etc.

    Δ_1 = Δ for all loops that do not touch forward path 1.
    Δ_2 = Δ for all loops that do not touch forward path 2.
    Δ_3 = etc.

This formula may seem difficult to use but only requires practice in order to master it.

8.   Find the transfer functions [Y(s)/X(s)] associated with figs 5.25a, b and c.

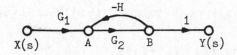

Fig.5.25a

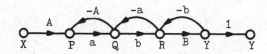

Fig.5.25b

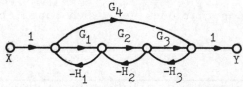

Fig.5.25c

    (a) There is only one forward path from X to Y and its total gain is G_1G_2, i.e. $M_1 = G_1G_2$, $M_2 = 0$, $M_3 = 0$, etc.

    There is only one loop and its gain is - G_2H, therefore

    Δ = 1 + G_2H + 0 + 0, etc.

    There are no loops that do not touch the forward path, therefore

    Δ_1 = 1, Δ_2 = 1, etc.

and $\dfrac{Y(s)}{X(s)} = \dfrac{G_1G_2}{1 + G_2H} \cdot 1$

(b) One forward path, therefore M_1 = AabB, M_2 = 0, etc.

$$\Delta = 1 - (- aA - ba - bB) + (AabB) + 0, \text{ etc.}$$

(2 non-touching loops)

There are no loops that do not touch the forward path, therefore

$$\Delta_1 = 1, \text{ etc.}$$

and $\dfrac{Y}{X}(s) = \dfrac{AabB}{1 + aA + ba + bB + AabB}$

$$= \left[1 + \frac{1}{Aa} + \frac{1}{bB} + \frac{1}{AB} + \frac{1}{abAB} \right]^{-1}$$

This method is obviously much quicker than that used in example 7.

(c) Two forward paths, therefore

$M_1 = G_4$ call this f.p.1
$M_2 = G_1G_2G_3$ call this f.p.2

$$\Delta = 1 + (G_1H_1 + G_2H_2 + G_3H_3 + G_4H_1H_2H_3) + (G_1H_1G_3H_3) + 0, \text{ etc.}$$

(2 non-touching loops)

There are no loops that do not touch f.p.2, so Δ_2 = 1.

There is one loop that does not touch f.p.1, therefore

$$\Delta_1 = 1 + G_2H_2 + 0, \text{ etc.}$$

and $\dfrac{Y}{X}(s) = \dfrac{G_4(1 + G_2H_2) + G_1G_2G_3}{1 + (G_1H_1 + G_2H_2 + G_3H_3 + G_4H_1H_2H_3) + G_1H_1G_3H_2}$

FURTHER EXAMPLES

1. A pump and motor flow control transmission system drives a load consisting of inertia J and viscous friction torque f per unit angular velocity. Draw a block diagram for the system and hence show that the transfer function is

$$\frac{\omega_m C_m}{\omega_p C_p}(s) = \frac{C_m^2}{2\lambda(Js + f) + C_m^2}$$

Neglect pipe friction and compressibility effects.

ω_m = motor speed C_m = motor capacity/rad

ω_p = pump speed C_p = pump capacity/rad

pump leakage coefficient = motor leakage coefficient
= λ which is a volumetric
rate per unit pressure rise

2. Using example 1 draw the signal flow diagram and establish the system transfer function from this.

3. Fig. 5.26 shows a transmission system where leakage, pipe friction and fluid compressibility are all considered.

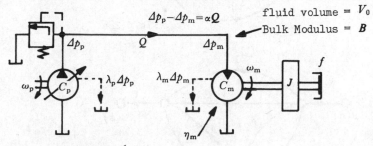

Fig.5.26

Show that the signal flow diagram is of the form shown in fig. 5.27.   Determine the values of a, b, τ_1 and τ_2 and derive the block diagram having one forward path and one feedback path only.

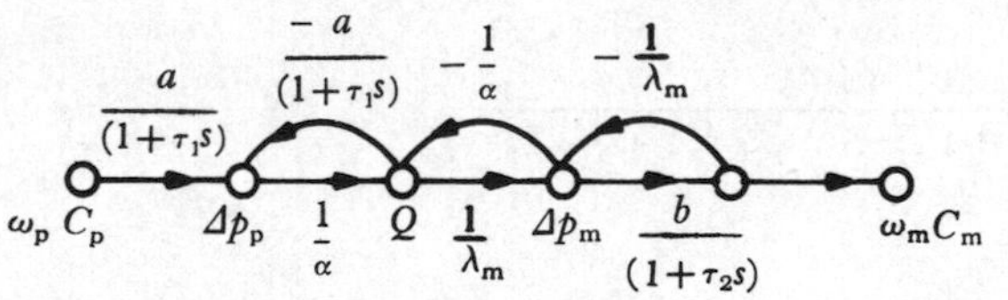

Fig.5.27

$[a = 1/\lambda_p, \ \tau_1 = V_0/(B\lambda_p), \ b = C_m^2 f/\eta_m, \ \tau_2 = J/f]$

4. Show that the transfer function obtained from the signal flow diagram of example 3 can be expressed in the form

$$\frac{\omega_m C_m}{\omega_p C_p}(s) = \frac{A \omega_n^2}{s^2 + 2\zeta\omega_n s + \omega_n^2}$$

where A is a constant
     ζ is the damping ratio
     ω_n is the natural undamped frequency

Show that

$$\omega_n^2 = \frac{1}{\tau_1\tau_2} \left(1 + ab + \frac{\alpha b}{1 + \alpha\lambda_m} \right)$$

161

5. Establish that the transfer functions obtained from signal flow diagrams figs 5.28a and b may be expressed as indicated.

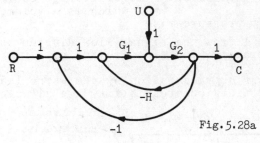

Fig.5.28a

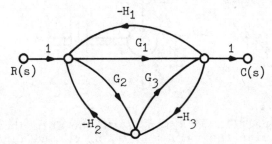

Fig.5.28b

(a) With $R(s) = 0$

$$\frac{C}{U}(s) = \frac{G_2(s)}{1 + G_1G_2(s) + G_1G_2H(s)}$$

(b)

$$\frac{C}{R}(s) = \frac{G_1 + G_2G_3}{1 + G_1H_1 + G_2H_2 + G_3H_3 + G_1G_3H_1 + G_1H_2H_3}$$

APPLICATION TO NON-LINEAR SYSTEM

As previously stated, only linear systems can be represented by block or signal flow diagrams unless the system can be linearised. This linearisation can take the form of a consideration of small perturbations about a particular operating point.

Consider the equation $y = kx^a z^b$. Now

$$\delta y = \frac{\partial y}{\partial x} \delta x + \frac{\partial y}{\partial z} \delta z$$

$$= kz^b a \frac{x^a}{x} \delta x + kx^a b \frac{z^b}{z} \delta z$$

$$= ya \frac{\delta x}{x} + yb \frac{\delta z}{z}$$

162

therefore

$$\delta y = \frac{ya}{x} \delta x + \frac{yb}{z} \delta z$$

or

$$\frac{\delta y}{y} = a \frac{\delta x}{x} + b \frac{\delta z}{z}$$

where δx, δy and δz are small perturbations about a particular point y having particular values x and z. Hence a linear relationship is obtained that is applicable close to a particular operating point. By such linearisation a transfer function may be obtained in terms of δy, δx, δz, etc., which has as its initial zero conditions the particular operating point i.e. $\delta y = \delta z = \delta z = 0$. A differential equation may be linearised in a similar manner, thus

$$M\overset{..}{x} + f\overset{.}{x} + x^2 = y$$

$$Ms^2 \delta X + fs \delta X + 2X \delta X = \delta Y$$

FURTHER EXAMPLES

6. Show that the linearised versions of the following equations are as indicated.

(a) $\quad Q = C_d bx \sqrt{\left(\frac{2 \Delta p}{\rho}\right)} \quad : \quad \frac{\delta Q}{Q} = \frac{\delta x}{x} + \frac{1}{2} \frac{\delta \Delta p}{p}$

(b) $\quad Q = kx \sqrt{\left(\frac{p_1 - p}{\rho}\right)} \quad : \quad \frac{\delta Q}{Q} = \frac{\delta x}{x} - \frac{1}{2} \frac{\delta p}{p_1 - p}$

p_1 = constant.

(c) $\quad Q = kx \sqrt{\left(\frac{p_1 - p_2}{\rho}\right)} \quad : \quad \frac{\delta Q}{Q} = \frac{\delta x}{x} + \frac{1}{2} \frac{\delta p_1 - \delta p_2}{p_1 - p_2}$

(d) $\quad T = J\overset{.}{\omega} + f\omega \quad\quad : \quad \delta T = (Js + f) \delta \omega(s)$

and if

$$T = (Js + f)\omega(s) \quad : \quad \frac{\delta T}{T} = \frac{\delta \omega}{\omega}$$

7. Show that the linearised equation for each component is as indicated.

(a) Variable-capacity pump delivering at high pressure (fig. 5.29).

$$\delta Q = \omega_p \delta C_p + C_p \delta \omega_p - \left[\lambda + \frac{V_0}{B} s\right] \delta p(s)$$

163

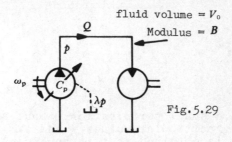

fluid volume $= V_0$

Modulus $= B$

Fig.5.29

(b) Relief valve - simple case (fig. 5.30).

$$\delta Q_R = \frac{R}{2}\left(\sqrt{p}\right)\left[3 - \frac{p_R}{p}\right]\delta p$$

Given that

$$Q_R = R(p - p_R)\sqrt{p} \quad \text{for } p \geqq p_R$$

(c) Variable-area orifice (fig. 5.31).

$$\frac{\delta Q}{Q} = \frac{\delta a}{a} + \frac{1}{2}\frac{\delta p_1 - \delta p_2}{p_1 - p_2}$$

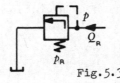

Fig.5.30

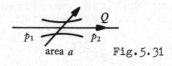

area a　　Fig.5.31

(d) Pipe flow (fig. 5.32).

Laminar $\quad \dfrac{\delta Q}{Q} = \dfrac{\delta p_1 - \delta p_2}{p_1 - p_2}$

Turbulent $\quad \dfrac{\delta Q}{Q} = \dfrac{1}{2}\dfrac{\delta p_1 - \delta p_2}{p_1 - p_2}$

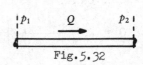

Fig.5.32

Given that

$$Q = c(p_1 - p_2) \quad \text{laminar}$$

$$Q = c\sqrt{(p_1 - p_2)} \quad \text{turbulent}$$

(e) Fluid acceleration with friction (fig. 5.33).

$$\delta Q = (\delta p_1 - \delta p_2)\left[\frac{m}{a^2}s + \frac{1}{c}\right]^{-1}$$

(f) Variable-capacity motor (fig. 5.34).

164

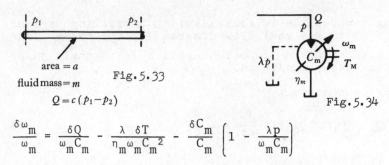

area = a

fluid mass = m

Fig. 5.33

$Q = c(p_1 - p_2)$

Fig. 5.34

$$\frac{\delta\omega_m}{\omega_m} = \frac{\delta Q}{\omega_m C_m} - \frac{\lambda}{\eta_m} \frac{\delta T}{\omega_m C_m^2} - \frac{\delta C_m}{C_m}\left(1 - \frac{\lambda p}{\omega_m C_m}\right)$$

(g)   Inertia + viscous + constant load (fig. 5.35).

$\delta T = (Js + b)\delta\omega$

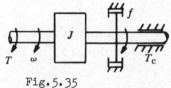

Fig. 5.35

WORKED EXAMPLES

9.   The diagrams (fig. 5.36) show relief valves where a constant pump delivery Q_p gives rise to a relief valve flow Q_R and a load flow Q_L when the relief valve is open. Since the valves are direct acting types the initial

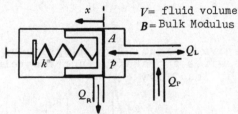

V = fluid volume

B = Bulk Modulus

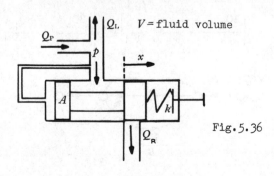

V = fluid volume

Fig. 5.36

165

spring compression is equivalent to a cracking pressure of p_0. Draw a signal flow diagram for the valves and obtain the transfer function applicable to small changes about a steady condition.

System equations

$$Q_p = Q_L + Q_R + A \frac{dx}{dt} + \frac{V}{B} \frac{dp}{dt} \qquad (5.1)$$

$$Q_R = bx\sqrt{p}$$

where b = port width. $\qquad (5.2)$

$$(p - p_0) A = kx + M \frac{d^2x}{dt^2} + f \frac{dx}{dt} \qquad (5.3)$$

where M = spool mass and f = damping coefficient. Linearisation gives (in the s domain)

$$\delta Q_p = \delta Q_L + \delta Q_R + A s \, \delta x + \frac{V}{B} s \, \delta p$$

But if Q_p = constant, δQ_p = 0, therefore

$$\delta Q_R + A s \, \delta x + \frac{V}{B} s \, \delta p = - \delta Q_L \qquad (5.1a)$$

and $\quad \dfrac{\delta Q_R}{Q_R} = \dfrac{\delta x}{x} + \dfrac{1}{2} \dfrac{\delta p}{p} \qquad (5.2a)$

and $\quad \delta p A = k \, \delta x + M s^2 \, \delta x + f s \, \delta x \qquad (5.3a)$

These equations give a signal flow diagram indicated in fig. 5.37. This gives a transfer function relating δp to small decreases in load flow δQ_L as

$$\frac{\delta p}{- \delta Q_L}(s) = \frac{B}{V}\left(s^2 + \frac{f}{M} s + \frac{k}{M}\right) \Bigg/ \Bigg[s^3 + \left(\frac{BQ_R}{2pV} + \frac{f}{M}\right)s^2 +$$

$$\left(\frac{BQ_R}{2pV} \frac{f}{M} + \frac{k}{M} + \frac{BA^2}{VM}\right)s + \frac{BQ_R A}{MVx} + \frac{BQ_R A}{2pVM}\Bigg]$$

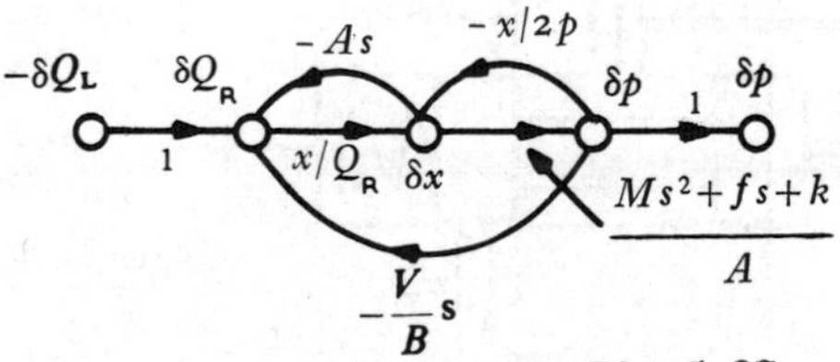

Fig.5.37

166

10. (Introduction to example 11). The measuring device indicated in fig. 5.38 enables the pressure p_2 to indicate the gap width x. The sensitivity is defined as the change of p_2 with x i.e. dp_2/dx. Obtain the conditions for maximum sensitivity and plot a dimensionless graph relating p_2 to x.

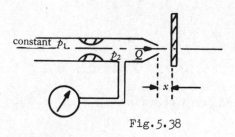

Fig.5.38

For incompressible flow

$$Q = k_1\sqrt{(p_1 - p_2)} = k_2\sqrt{(p_2)} \; x$$

where $k_1 = c_{d_1} a_1\sqrt{(2/\rho)}$

$\quad\quad\quad k_2 = C_{d_2} b_2\sqrt{(2/\rho)}$

$\quad\quad\quad C_d$ = coefficient of discharge

$\quad\quad\quad a_1$ = orifice area of restriction

$\quad\quad\quad b$ = orifice perimeter of nozzle

Therefore

$$k_1{}^2(p_1 - p_2) = k_2{}^2 \; x^2 p_2$$

i.e. $\dfrac{p_2}{p_1} = \dfrac{k_1{}^2}{k_1{}^2 + x^2 k_2{}^2} = \dfrac{1}{1 + \left(\dfrac{k_2}{k_1} x\right)^2}$

thus $\dfrac{dp_2}{dx} = p_1\left\{- 2\left(\dfrac{k_2}{k_1}\right)^2 x \left/ \left[1 + \left(\dfrac{k_2}{k_1} x\right)^2\right]^2\right.\right\}$

For maximum sensitivity

$$\dfrac{d\left(\dfrac{dp_2}{dx}\right)}{dx} = 0$$

hence

$$\left(\dfrac{k_2}{k_1} x\right)^2 = \dfrac{1}{3}$$

Graph (fig. 5.39). Using

167

$$\frac{p_2}{p_1} = \frac{1}{1 + \left(\dfrac{k_2}{k_1} x\right)^2}$$

and putting

$$\frac{k_2}{k_1} x = 0 \ 0.2 \ 0.4 \ 0.6 \ 0.8 \ 1.0 \ 2 \ 3 \ 4 \ 5$$

gives

$$\frac{p_2}{p_1} = 1 \ 0.962 \ 0.862 \ 0.735 \ 0.610 \ 0.5 \ 0.2 \ 0.1 \ 0.059 \ 0.038$$

The graph is plotted as indicated in fig. 5.39.

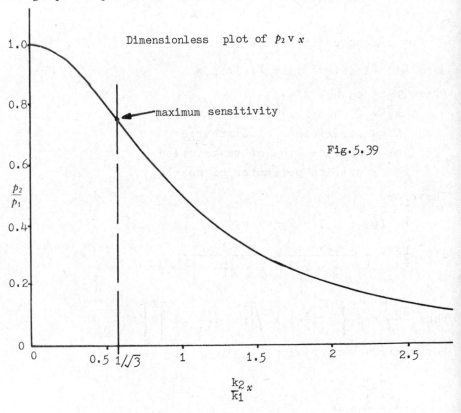

11. A development of example 10 is the flapper nozzle system where variations of the gap x are used to determine the motion of a load as indicated in fig. 5.40. Determine the transfer function $\delta y/\delta x$ when p_1 is constant and the transfer function $\delta y/\delta p_1$ when δx is fixed.

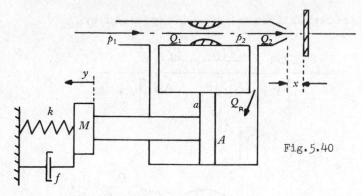

Fig. 5.40

$$Q_1 = k_1 \sqrt{(p_1 - p_2)}$$

thus $\dfrac{\delta Q_1}{Q_1} = \dfrac{1}{2} \dfrac{\delta p_1 - \delta p_2}{p_1 - p_2}$

$$Q_2 = k_2 x \sqrt{p_2}$$

thus $\dfrac{\delta Q_2}{Q_2} = \dfrac{\delta x}{x} + \dfrac{1}{2} \dfrac{\delta p_2}{p_2}$

$$Q_R = Q_1 - Q_2 = A \dfrac{dy}{dt}$$

thus $\delta Q_1 - \delta Q_2 = As\, \delta y$

$$p_2 A - p_1 a = M \dfrac{d^2 y}{dt^2} + f \dfrac{dy}{dt} + ky$$

thus $\delta p_2 A - \delta p_1 a = (Ms^2 + fs + k)\, \delta y$

(in the s domain). These equations give

$$\frac{Q_1}{2(p_1 - p_2)}\, \delta p_1 - \frac{1}{2}\left[\frac{Q_1}{p_1 - p_2} + \frac{Q_2}{p_2}\right]\delta p_2 - \frac{Q_2}{x}\, \delta x = As\, \delta y$$

and $\quad \delta p_2 A - \delta p_1 a = (Ms^2 + fs + k)\, \delta y$

both in the s domain even though lower-case letters are used. From these equations a signal flow diagram may be drawn as shown in fig. 5.41.

With p_1 = constant, $\delta p_1 = 0$, therefore

$$\frac{\delta y}{\delta x}(s) = \frac{-\dfrac{2Q_2}{x}\left[1 + \dfrac{Q_2}{Q_1}\left(\dfrac{p_1}{p_2} - 1\right)\right]}{Ms^2 + \left[f + \dfrac{2A^2}{1 + \dfrac{Q_2}{Q_1}\left(\dfrac{p_1}{p_2} - 1\right)}\right]s + k}$$

169

The effect of variations of p_1 is expressed by

$$\frac{\delta y}{\delta p_1}(s) = \frac{ZA\left[1 + \dfrac{2a(p_1 - p_2)}{Q_1}\, s\right]}{Ms^2 + \left[f + \dfrac{2A^2Z(p_1 - p_2)}{Q_1}\right]s + k}$$

where

$$Z = \left[1 + \frac{Q_2}{Q_1}\left(\frac{p_1}{p_2} - 1\right)\right]^{-1}$$

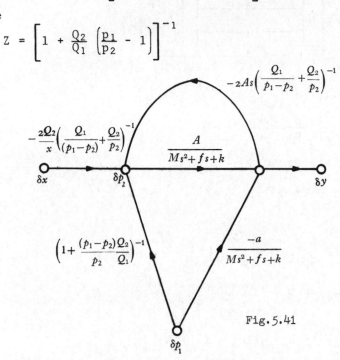

Fig. 5.41

12.   Fig. 5.42 shows a fixed capacity pump delivering oil to a simple bleed-off speed control system. The pump delivery Q_p is constant and the pump pressure p_1 never

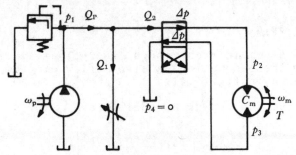

Fig. 5.42

reaches the relief valve setting. The flow through the variable-area orifice is given by $Q_1 = k_1A\sqrt{p_1}$ where A is

170

the orifice area. The flow through each path of the 4/2 valve is $Q_2 = k_2\sqrt{\Delta p}$ where Δp is the associated path pressure drop. At the motor $Q_2 = C_m\omega_m$ where C_m = motor capacity per radian and ω_m = motor speed rad/s. The motor torque $T = J\dot{\omega}_m + c\omega_m$ where J = motor inertia and c = viscous load torque per unit speed.

(a) Determine the transfer function relating small changes in motor speed to small changes in valve area $(\delta\omega_m/-\delta A)$.

(b) Determine the transfer function when the motor torque is made constant.

(a)

$$Q_p = Q_1 + Q_2$$

thus $\delta Q_1 = -\delta Q_2$ since $\delta Q_p = 0$

$$Q_1 = k_1 A\sqrt{p_1}$$

thus $\dfrac{\delta Q_1}{Q_1} = \dfrac{\delta A}{A} + \dfrac{1}{2}\dfrac{\delta p_1}{p_1}$

$$Q_2 = k_2\sqrt{p_3}$$

thus $\dfrac{\delta Q_2}{Q_2} = \dfrac{1}{2}\dfrac{\delta p_3}{p_3}$

Also $Q_2 = \omega_m C_m$

thus $\dfrac{\delta Q_2}{Q_2} = \dfrac{\delta\omega_m}{\omega_m}$

$$(p_2 - p_3) C_m = T$$

thus $\delta T = C_m (\delta p_2 - \delta p_3)$

and $T = (Js + c)\omega_m$

thus $\delta T = (Js + c)\delta\omega_m$

$$p_1 - p_2 = p_3$$

thus $\delta p_1 - \delta p_2 = \delta p_3$

These equations will give the following signal flow diagram from which the transfer function $\delta\omega_m/\delta A$ is obtained (fig. 5.43).

171

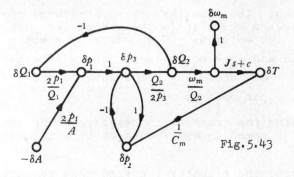

Fig.5.43

$$\frac{\delta\omega_m}{-\delta A}(s) = \frac{2p_1Q_1Q_2/A}{\omega_mQ_1(Js + c) + 2C_m(2p_3Q_1 + p_1Q_2)}$$

(b) For a constant motor torque T = constant, therefore $\delta T = 0$, therefore

$$\delta p_2 = \delta p_3$$

and since $\delta p_1 - \delta p_2 = \delta p_3$, then

$$\delta p_1 = 2\delta p_3 = 2\delta p_2$$

This gives fig. 5.44, and

$$\frac{\delta\omega_m}{-\delta A}(s) = \frac{p_1Q_1\omega_m/A}{2p_3Q_1 + p_1Q_2}$$

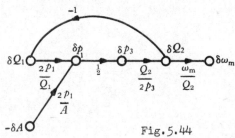

Fig.5.44

13. A valve controlled transmission system is indicated
in fig. 5.45 and the pump delivery may be considered constant at Q_p. The direct-acting relief valve flow is given
by the expression $Q_R = k_R(p - p_R)\sqrt{p}$ where p is the pump
delivery pressure and p_R is the relief valve setting. The
5/2 metering and directional valve has a flow rate through
each flow path given by $Q_L = k_L x\sqrt{(2\Delta p)}$ where x is the valve
spool displacement and Δp is the flow path pressure drop.

The motor has a capacity of C_m per radian and volume-

172

tric and mechanical efficiencies of η_v and η_m which may be considered as constant at 100%. The motor load consists of an inertia J and a viscous torque of f per unit angular velocity.

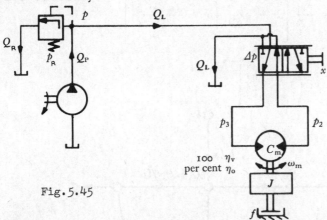

Fig. 5.45

Determine the transfer function relating changes of motor speed to small changes of spool position relative to a particular operating point.

System equations	linearised form

Q_p = constant

$$\delta Q_p = 0$$

$Q_L = Q_p - Q_R$

$$\delta Q_L = -\,\delta Q_R$$

$Q_R = k_R(p - p_R)\sqrt{p}$

$$\delta Q_R = \frac{k_R}{2}\sqrt{p}\left[3 - \frac{p_R}{p}\right]\delta p$$

$Q_L = k_L x\sqrt{(2\Delta p)}$

$$\frac{\delta Q_L}{Q_L} = \frac{\delta x}{x} + \frac{1}{2}\frac{\delta \Delta p}{\Delta p}$$

$Q_L = \omega_m C_m$

$$\frac{\delta Q_L}{Q_L} = \frac{\delta \omega_m}{\omega_m}$$

$p - 2\Delta p = \Delta p_L$

$$\delta p - 2\delta\Delta p = \delta\Delta p_L$$

(where $\Delta p_L = p_2 - p_3$)

$T_m = \Delta p_L\, C_m$

$$\frac{\delta T_m}{T_m} = \frac{\delta\Delta p_L}{\Delta p_L}$$

(where T_m = motor torque)

$T_m = J\dot{\omega}_m + f\omega_m$

$$\delta T_m = (Js + f)\delta\omega_m$$

(in s domain)

173

These equations will give a signal flow diagram which may be drawn as in fig. 5.46. This gives the transfer function

$$\frac{\delta \omega_m}{\delta x}(s) = \frac{\omega_m}{x}\left[1 + \frac{Q_L}{4\Delta p}\left(a + \frac{Js + f}{C_m{}^2}\right)\right]^{-1}$$

where

$$a = 2\left[k_R\sqrt{p}\left(3 - \frac{p_R}{p}\right)\right]^{-1}$$

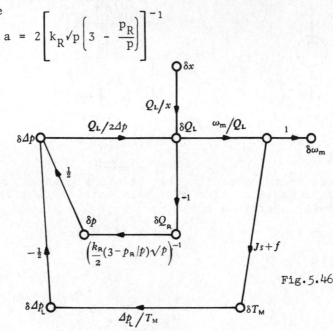

Fig.5.46

FURTHER EXAMPLES

8. Fig. 5.47 shows a direct-acting relief valve with a damping orifice through which the flow rate is given by $q = c(p - p_2)$. Show that the transfer function, derived from a signal flow diagram, is

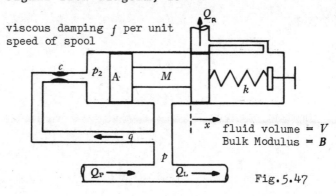

viscous damping f per unit speed of spool

fluid volume = V
Bulk Modulus = B

Fig.5.47

174

$$\frac{\delta p}{-\delta Q_L}(s) = \frac{B}{V}\left[s^2 + \left(\frac{f}{M} + \frac{A^2}{cM}\right)s + \frac{k}{M}\right]\Bigg/\left[s^3 + \right.$$

$$\left(\frac{BQ_R}{2pV} + \frac{f}{M} + \frac{A^2}{cM}\right)s^2 +$$

$$\left.\left(\frac{BQ_Rf}{2pVM} + \frac{k}{M} + \frac{BA^2}{VM} + \frac{BA^2Q_R}{2pVcM}\right)s + \frac{BAQ_R}{MVx}\right]$$

9. Fig. 5.48 shows a transmission system where the following data apply. Q_p = constant, $Q_R = k_R(p - p_R)\sqrt{p}$, $Q_L = k_L x\sqrt{(2\Delta p)} = \omega_m C_m$, motor load torque T_m is independent of speed. Construct a signal flow diagram indicating the relationship between $\delta\omega_m$, δx and δT_m.

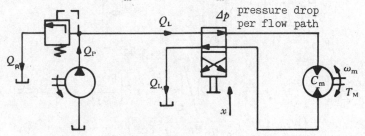

Fig. 5.48

When T_m is constant, show that

$$\frac{\delta\omega_m}{\delta x}(s) = \frac{\omega_m}{x}\left[1 + \frac{a(k_L x)^2}{2\omega_m C_m}\right]^{-1}$$

and when x is constant, show that

$$\frac{\delta\omega_m}{\delta x}(s) = -\left(a + \frac{4\Delta p}{Q_L}\right)^{-1}$$

where

$$a = 2\left[k_R\sqrt{p}\left(3 - \frac{p_R}{p}\right)\right]^{-1}$$

10. The pilot relief valve shown in fig. 5.49 has an orifice flow $q = c(p_1 - p_2)$. This flow may be assumed equal to the flow through the pilot spool and p_2 may be assumed constant when the relief valve is passing oil to tank (Q_R).

The flow through the main spool (Q_{R_1}) is given by

$$Q_{R_1} = k_1 x\sqrt{p_1}$$

175

The total oil volume is V and the bulk modulus is B. Show that the transfer function relating small changes of pressure (δp_1) to flow changes ($-\delta Q_L$) about a fixed operating point is

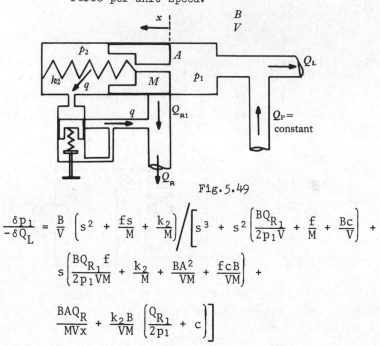

f = viscous damping
force per unit speed.

Fig. 5.49

$$\frac{\delta p_1}{-\delta Q_L} = \frac{B}{V}\left[s^2 + \frac{fs}{M} + \frac{k_2}{M}\right]\Bigg/\Bigg[s^3 + s^2\left(\frac{BQ_{R_1}}{2p_1 V} + \frac{f}{M} + \frac{Bc}{V}\right) +$$

$$s\left(\frac{BQ_{R_1}}{2p_1 VM}f + \frac{k_2}{M} + \frac{BA^2}{VM} + \frac{fcB}{VM}\right) +$$

$$\frac{BAQ_R}{MVx} + \frac{k_2 B}{VM}\left(\frac{Q_{R_1}}{2p_1} + c\right)\Bigg]$$

Also show that the signal flow diagram may be drawn as in fig. 5.50.

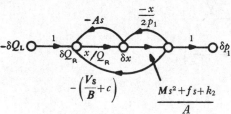

Fig. 5.50

11. The circuit of fig. 5.51 indicates a simple bleed-off speed control system. The pump and motor leakage rate may each be expressed as λp_1. The relief valve flow is given by $Q_R = R(p_1 - p_R)\sqrt{p_1}$. The bleed-off flow is given by $Q_B = k_B a_0 \sqrt{p_1}$. The motor torque is given by $T = J\dot\omega_m + f\omega_m$. All mechanical efficiencies may be taken as

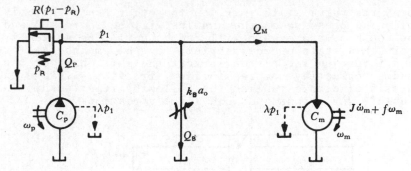

Fig.5.51

100%. If compressibility is neglected, show that one form of the signal flow diagram, relating small changes of the variables about an operating point, is as indicated in fig. 5.52.

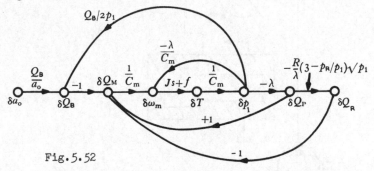

Fig.5.52

12. The simple circuit shown in fig. 5.53 is employed where the piston load force F is likely to be either positive or negative during motion from left to right.

The primary feed control is a variable-delivery pump which transfers oil from one end of the two-rodded piston to the other. Provided the pump is free from internal leakage the feed rate is independent of oil viscosity and the boost pump ensures the system is at a sufficient pressure to prevent sponginess.

The function of the valve is to keep the pressure difference between the two sides of the piston constant whatever the pull or push from the work, etc. If there is no load on the piston then the pressures in both ends of the cylinder must be the same and the valve acts as an ordinary relief valve, the blow-off pressure depending on the strength of the spring and the area of the valve spool head. A resistant load will tend to reduce the downstream pressure, reducing the pressure on the spool annulus. The boost of pump pressure will, therefore, increase until it opens the valve again. If, on the other hand, there is a pull, the pressure on the annulus will be increased and the boost pump pressure will reduce until the system is again in a state of equilibrium. By this means the forward thrust on the piston is kept the same whatever the load. Neglecting compressibility, friction, leakage and inertia effects establish the following relationship for pressure variations and load force variations

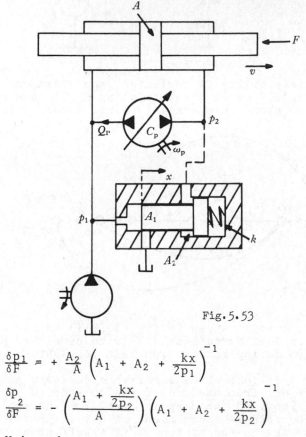

Fig. 5.53

$$\frac{\delta p_1}{\delta F} = + \frac{A_2}{A} \left(A_1 + A_2 + \frac{kx}{2p_1} \right)^{-1}$$

$$\frac{\delta p_2}{\delta F} = - \left(\frac{A_1 + \frac{kx}{2p_2}}{A} \right) \left(A_1 + A_2 + \frac{kx}{2p_2} \right)^{-1}$$

13. Using the same system as in question 12, assume that the load $F = M(dv/dt) + fv$, where M = load mass and f = viscous damping force per unit velocity, and that the variable capacity pump leakage is given by Q(leakage to tank) = λp_1. Establish the transfer function

$$\frac{\delta v}{\delta Q_p} = A \left(1 + \frac{\lambda kx/(Ms + f)}{2AA_2 p_1} \right)^{-1}$$

where pump swept volume $Q_p = \omega_p C_p$.

178

6 FEEDBACK SYSTEMS

All the systems considered in the previous sections have
been concerned with the control of: (a) speed, (b) pressure and (c) direction. Many of the systems had inherent
feedback loops due to leakage or compressibility, but such
feedback did not directly assist the control system in per-
forming its function. Some of the systems, e.g. pressure
compensated speed control valve, had deliberate feedback
loops without which the control would not have been possible. Further systems will now be considered which involve deliberate feedback loops and consideration will be
given to position control.

PRESSURE CONTROL

The control of system primary pressures by a relief valve
- either direct acting or pilot type in parallel with the
pump - has been considered previously and the relevant
transfer functions were established.

Control of Secondary Pressure by Reducing Valve

Again this may be done by either a direct acting or pilot
type valve. An analysis of a direct acting type is considered where it should be noted that the valve is in
series with the pump, and p_1 is frequently assumed constant while p_2 is set by the reducing valve.

Consider the system indicated by fig. 6.1.

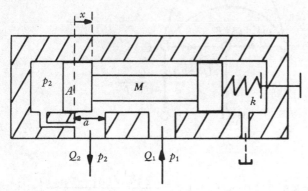

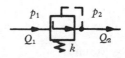

Fig.6.1

System equations

$$Q_2 = Q_1 - A \frac{dx}{dt}$$

$$Q_1 = b(a - x)\sqrt{(p_1 - p_2)}$$

where a = max. valve port opening when x = 0
      b = port flow factor

$$p_2 A = M\ddot{x} + c\dot{x} + k(x_0 + x)$$

where c = viscous damping coefficient for spool
      x_0 = initial compression of spool spring

For small perturbations (zero initial conditions)

$$\delta Q_2 = \delta Q_1 - As\,\delta x$$

$$\frac{\delta Q_1}{Q_1} = -\frac{\delta x}{a - x} + \frac{1}{2}\frac{\delta p_1 - \delta p_2}{p_1 - p_2}$$

$$A\,\delta p_2 = (Ms^2 + cs + k)\,\delta x$$

These equations give a signal flow diagram as indicated
in fig. 6.2. This shows that δp_2 is a function of δQ_2
and δp_1 and the following transfer functions may be obtained.

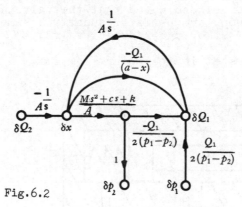

Fig.6.2

If $\delta Q_2 = 0$ then

$$\frac{\delta p_2}{\delta p_1} = (Ms^2 + cs + k)\bigg/\left\{Ms^2 + \left[c + \frac{2A^2}{b}\frac{\sqrt{(p_1 - p_2)}}{(a - x)}\right]s + \left[k + \frac{2A(p_1 - p_2)}{a - x}\right]\right\}$$

If $\delta p_1 = 0$ then

180

$$\frac{\delta p_2}{\delta Q_2} = - \frac{2\sqrt{(p_1 - p_2)}}{b(a - x)} \; (Ms^2 + cs + k) \Big/ \Big\{ Ms^2 +$$

$$\left[c + \frac{2A^2\sqrt{(p_1 - p_2)}}{b(a - x)} \right] s + \left[k + \frac{2A(p_1 - p_2)}{a - x} \right] \Big\}$$

For conditions where M, c and $- A\overset{\prime}{x}$ can be neglected it is seen that

$$\frac{\delta p_2}{\delta p_1} = \frac{1}{1 + \frac{2A}{k}\left(\frac{p_1 - p_2}{a - x}\right)}$$

Also $\dfrac{\delta p_2}{\delta Q_2} = \dfrac{1}{\dfrac{Q_2}{2(p_1 - p_2)} + \dfrac{AQ_2}{k(a - x)}} \approx \dfrac{2(p_1 - p_2)}{Q_2}$

With pneumatic systems the secondary pressure is obtained by use of a pressure-reducing valve, more frequently known as a pressure-regulating valve, where the secondary pressure is also maintained at zero flow. Fig. 6.3 illustrates this valve.

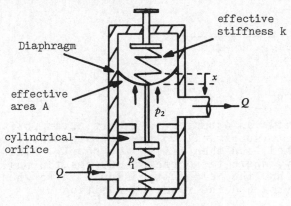

Fig.6.3

Let pre-compression of regulating spring be equivalent to a pressure p_R acting on area A. Neglecting the area of the valve poppet and assuming a mass M and damping coefficient c for the assembly, then

$$(p_2 - p_R) A - kx = M\overset{\prime\prime}{x} + c\overset{\prime}{x}$$

$$Q = bx\sqrt{(p_1 - p_2)}$$

where b = orifice flow factor; therefore

$$\frac{\delta Q}{Q} = \frac{\delta x}{x} + \frac{1}{2(p_1 - p_2)} \; (\delta p_1 - \delta p_2)$$

181

and $A \delta p_2 = (Ms^2 + cs + k) \delta x$

therefore

$$\delta p_2 \left[\frac{Ax^{-1}}{Ms^2 + cs + k} - \frac{1}{2(p_1 - p_2)} \right] = \frac{\delta Q}{Q} - \frac{1}{2(p_1 - p_2)} \delta p_1$$

and if M and c are ignored then

$$\delta p_2 \left[\frac{A}{kx} - \frac{1}{2(p_1 - p_2)} \right] = \frac{\delta Q}{Q} - \frac{1}{2(p_1 - p_2)} \delta p_1$$

If $\delta p_1 = 0$, then

$$\frac{\delta p_2}{\delta Q} = \frac{1}{Q \left[\frac{A}{kx} - \frac{1}{2(p_1 - p_2)} \right]}$$

$$\simeq \frac{kx}{AQ}$$

If $\delta Q = 0$, then

$$\frac{\delta p_2}{\delta p_1} = \frac{-1}{\frac{2A}{kx}(p_1 - p_2) - 1}$$

POSITION CONTROL

In order to move an actuator to a particular position it is necessary to supply oil to the actuator until the position is reached, and then remove the supply. This may be done manually, where the operator watches the position of the actuator and then closes the supply valve when the actuator has reached the required position.

For repetitive, accurate control, a system can be built which continuously monitors particular parameters of the system and performs actions aimed at moving the actuator to the required position. Such a system is known as a feedback control system and has to be equipped to monitor the basic parameters that are being controlled. A simple feedback system for position control would need to monitor the desired position of the actuator (input or reference quantity) and its present position (output or controlled variable quantity). The error would then be determined where

error = output - input

and, for proportional control, a correcting signal, proportional to the error in magnitude, but opposite in sign, would be applied to the control element. Hence the correcting signal ∝ input - output.

The control element would drive the output until no correcting signal is received. Under ideal conditions, at zero correcting signal, input = output, i.e. the actuator has moved to the desired position. The block representation of this is shown in fig. 6.4.

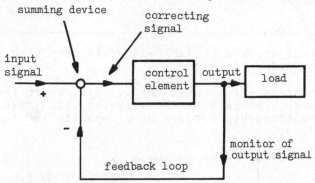

Fig.6.4

An example of this block would be as indicated in fig. 6.5. For such a system it is possible to establish the equation of motion and then study the theoretical response of the system.

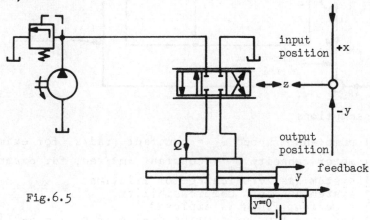

Fig.6.5

By inspection it can be seen that $Q \propto z$ provided loads, etc., remain constant. Now the actuator output speed = Q/A, where A = ram area. Therefore

$$\frac{dy}{dt} \propto z \quad \text{or} \quad \frac{dy}{dt} \propto x - y$$

Hence for large errors there is high-speed motion, and as

183

the error is reduced, so the correcting signal decreases and the output speed decreases until dy/dt = 0 at x = y.

It should be noted that as far as position (x, y) is concerned, the actuator is an integrating element i.e. y ∝ ∫(valve movement)dt and the provision of a feedback loop around an integrator is vital for stability. Feedback also tends to make a non-linear system more linear for small changes.

PUMP/MOTOR SYSTEMS

Position Control using a Variable-capacity Pump with a Fixed-capacity Hydraulic Motor

This arrangement is illustrated in fig. 6.6, where the pump stroke mechanism is moved by a system which results in a linear relationship between pump capacity and the correcting signal.

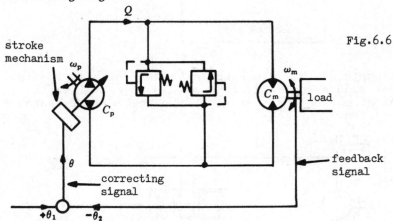

Fig.6.6

Assumptions

(a) pump shaft speed ω_p = constant (rad/s, for example)
(b) motor capacity C_m = constant (m³/rad, for example)
(c) system has zero leakage or friction
(d) system has zero compressibility
(e) a perfect fluid is employed
(f) response time of summing device ($\theta = \theta_1 - \theta_2$) is very short and can be neglected
(g) relief valve is not open
(h) $C_p = k_p\theta$ (where C_p is in m³/rad and θ in rad, for example)

Hence

$$Q = \omega_p C_p = \omega_p k_p \theta = \omega_p k_p (\theta_1 - \theta_2)$$

184

and $\quad Q = \omega_m C_m = C_m \dfrac{d\theta_2}{dt}$

thus $\quad \dfrac{C_m}{\omega_p k_p} \dfrac{d\theta_2}{dt} + \theta_2 = \theta_1$

$C_m/\omega_p k_p$ may be expressed as a time-constant τ_1 giving the differential equation for the system as

$$\tau_1 \dfrac{d\theta_2}{dt} + \theta_2 = \theta_1$$

or, for zero initial conditions, the transfer functions as

$$\dfrac{\theta_2}{\theta_1}(s) = \dfrac{1}{\tau_1 s + 1}$$

The solution (time-domain) for a step-input is of the form

$$\theta_2(t) = \theta_1(t)(1 - e^{-t/\tau_1})$$

which is represented by fig. 6.7.

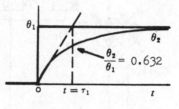

Fig.6.7

WORKED EXAMPLES

1. A hydraulic position control servo systém is illustrated in fig. 6.8.

The leakage rate from the pump is given by the expression $Q_{1p} = \lambda_p p$ and for the motor $Q_{1m} = \lambda_m p$, where λ is a constant. The effective volume of the high-pressure side is V_0 at atmospheric conditions and the effective bulk modulus is B. Neglecting friction and the response time of the summing device and stroke mechanism, establish the transfer function for the system.

Flow equation

$$Q = \omega_p C_p - \lambda_p p - \dfrac{V_0}{B} \dfrac{dp}{dt}$$

pump delivery   leakage   compressibility loss

185

and $Q = \omega_m C_m + \lambda_m p$

therefore

$$\omega_m = \frac{\omega_p}{C_m} C_p - \frac{(\lambda_p + \lambda_m)p}{C_m} - \frac{V_0}{C_m B} \frac{dp}{dt}$$

Torque equation: for 100% mechanical efficiency

$$pC_m = \text{Torque at motor shaft} = J \frac{d\omega_m}{dt}$$

and so $p = \dfrac{J}{C_m} \dfrac{d\omega_m}{dt}$

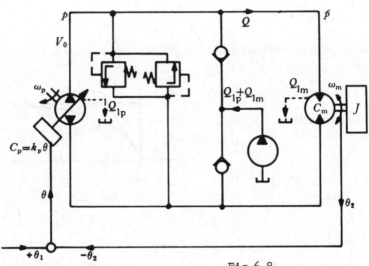

Fig.6.8

Hence

$$\omega_m + \frac{J(\lambda_p + \lambda_m)}{C_m{}^2} \frac{d\omega_m}{dt} + \frac{JV_0}{BC_m{}^2} \frac{d^2\omega_m}{dt^2} = \frac{\omega_p}{C_m} C_p$$

But $C_p = k_p \theta = k_p(\theta_1 - \theta_2)$

and $\omega_m = \dfrac{d\theta_2}{dt}$

Assuming zero initial conditions the Laplace form of the equation is

$$\frac{JV_0}{BC_m{}^2} s^3 \theta_2 + \frac{J(\lambda_p + \lambda_m)}{C_m{}^2} s^2 \theta_2 + s\theta_2 + \frac{\omega_p k_p}{C_m} \theta_2 = \frac{\omega_p k_p}{C_m} \theta_1$$

giving a transfer function of

$$\frac{\theta_2}{\theta_1}(s) = \frac{1}{\dfrac{JV_0}{BC_m\omega_p k_p}s^3 + \dfrac{J(\lambda_p + \lambda_m)}{C_m\omega_p k_p}s^2 + \dfrac{C_m}{\omega_p k_p}s + 1}$$

This transfer function may be written as

$$\frac{\theta_2}{\theta_1}(s) = \frac{1}{\dfrac{C_m}{\omega_p k_p}s\left[\dfrac{1}{\omega_n^2}s + \dfrac{2\zeta}{\omega_n}s + 1\right] + 1}$$

$$= \frac{1}{\dfrac{1}{K_M}s\left[\dfrac{1}{\omega_n^2}s + \dfrac{2\zeta}{\omega_n}s + 1\right] + 1}$$

where $K_M = \dfrac{\omega_p k_p}{C_m}$

$$\omega_n^2 = \frac{BC_m^2}{JV_0} \quad \text{i.e.} \quad \omega_n = \sqrt{\frac{K}{J}}$$

$$K = \text{system stiffness} = \frac{BC_m^2}{V_0}$$

$$\zeta = \frac{(\lambda_p + \lambda_m)}{2C_m}\sqrt{\frac{JB}{V_0}}$$

The reader should establish that the above transfer function is represented by the signal flow diagram in fig. 6.9, i.e. unity feedback around an integrator (pump/motor unit), with an oscillatory delay. K_M is the steady-state gain of the motor unit.

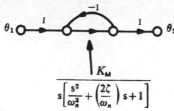

$$\frac{K_M}{s\left[\dfrac{s^2}{\omega_n^2} + \left(\dfrac{2\zeta}{\omega_n}\right)s + 1\right]}$$

Fig.6.9

Further, it can be seen that

(a) if compressibility is neglected the transfer function becomes

187

$$\frac{\theta_2}{\theta_1}(s) = \frac{1}{\tau_1 \tau \, s^2 + \tau_1 \, s + 1}$$

where

$$\tau_1 = \frac{C_m}{\omega_p k_p} = \frac{1}{K_M}$$

and

$$\tau = \frac{J}{C_m^2} (\lambda_p + \lambda_m)$$

or

$$\frac{\theta_2}{\theta_1}(s) = \frac{1}{\tau_1 \, s(\tau \, s + 1) + 1}$$

The reader should establish that this transfer function reveals a unity feedback around an integrator (pump/motor unit), with a single time delay as indicated by the signal flow diagram in fig. 6.10.

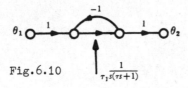

Fig.6.10

Note that

$$\frac{1}{\tau_1} = \frac{\omega_p k_p}{C_m} = K_M$$

where K_M is the steady state gain of the motor unit, and the loop gain of the system.

(b) if compressibility and leakage can be neglected the transfer function becomes

$$\frac{\theta_2}{\theta_1}(s) = \frac{1}{\tau_1 \, s + 1}$$

where

$$\tau_1 = \frac{C_m}{\omega_p k_p} = \frac{1}{K_M}$$

which is the same result as was obtained previously.

2. Construct the signal flow diagram for worked example 1.

The system equations are

$$C_p = k_p (\theta_1 - \theta_2)$$

$$Q = \omega_p C_p - \lambda_p p - \frac{V_0}{B} s \, p$$

and
$$Q = \omega_m C_m + \lambda_m p$$

$$p = \frac{J}{C_m} s \, \omega_m$$

$$\omega_m = s \, \theta_2$$

These equations lead to a signal flow diagram as shown in fig. 6.11. The deliberate feedback loop $(- k_p \theta_2)$ is

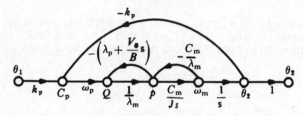

Fig.6.11

shown here together with the inherent feedback loops due to leakage and compressibility. The reader should establish the transfer function from this signal flow diagram.

Considering the Pump Stroke Mechanism

The force required to operate the stroke lever of a variable-capacity pump is frequently large and calls for some extra unit to provide the motion. A small linear actuator is often employed which is controlled via a 4/3 closed centre metering valve. Such a system is illustrated in fig. 6.12, but it will be shown that this is not a suitable arrangement.

$$x = k_d \theta = k_d (\theta_1 - \theta_2)$$

$Q_1 = K_x x$ for constant pressures and linear ports

$sy = \dfrac{Q_1}{A}$ in Laplace form

$\omega_m C_m = Q_2 = \omega_p C_p = \omega_p k_p y$ for no leakage or compressibility

$s \theta_2 = \omega_m = \dfrac{Q_2}{C_m}$

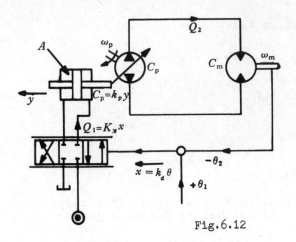

Fig.6.12

From these equations a signal flow diagram can be constructed as shown in fig. 6.13 (without the dotted line loop). From this the transfer function is obtained

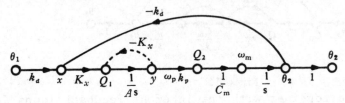

Fig.6.13

$$\frac{\theta_2}{\theta_1}(s) = \frac{\dfrac{k_d K_x k_p \omega_p}{AC_m}}{s^2 + \dfrac{k_d K_x k_p \omega_p}{AC_m}}$$

i.e. $\dfrac{\theta_2}{\theta_1}(s) = \dfrac{\omega_0{}^2}{s^2 + \omega_0{}^2}$

where $\omega_0{}^2 = k_d K_M \dfrac{K_x}{A} = k_d K_M K_L$

$\quad K_M$ = motor gain

$\quad K_L$ = actuator gain

$\quad k_d$ = differencing device gain

For a step input $\theta_1(t) = \theta_A$ i.e. $\theta_1(s) = \theta_A/s$. The solution for $\theta_2(t)$ is

$$\theta_2(t) = \theta_A[1 - \cos \omega_n t]$$

190

which is not acceptable for a position control system where the object is to make $\theta_2 = \theta_1$.

If the signal flow diagram of fig. 6.13 is reconsidered with an additional loop shown by the dotted line, then the transfer function becomes

$$\frac{\theta_2}{\theta_1}(s) = \frac{k_d K_x k_p \omega_p}{AC_m} \cdot \frac{1}{s^2 + \dfrac{k_d K_x k_p \omega_p}{AC_m} + \dfrac{K_x}{A} s}$$

i.e. $\dfrac{\theta_2}{\theta_1}(s) = \dfrac{\omega_0{}^2}{s^2 + 2\zeta\omega_0 s + \omega_0{}^2}$

i.e. a feedback loop has been put around the integrator (1/As). This is an acceptable system where the damping ratio ζ may be adjusted to a desired value. Introducing this loop means that

$$Q_1 = K_x(x - y)$$

instead of

$$Q_1 = K_x x$$

i.e. the valve port opening is the difference between the spool displacement and the actuator piston travel. This is achieved by connecting the 4/3 valve body to the actuator piston as illustrated by fig. 6.14a. Fig. 6.14b shows an alternative arrangement where the valve body is integral with the actuator cylinder. Here the actuator piston is fixed and the stroke mechanism is operated by the movement of the actuator cylinder.

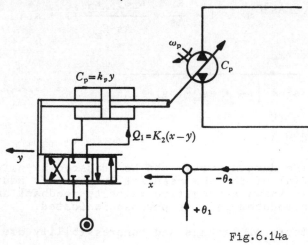

Fig.6.14a

191

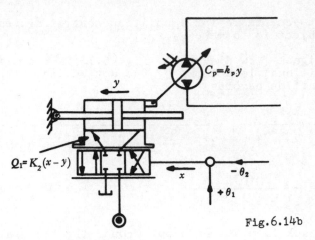

$C_p = k_p y$

$Q_1 = K_2(x - y)$

x $-\theta_2$

$+\theta_1$

Fig.6.14b

If fig. 6.13 is redrawn to include yet another feedback loop, as indicated in fig. 6.15, then the transfer function becomes

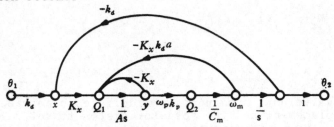

Fig.6.15

$$\frac{\theta_2}{\theta_1}(s) = \frac{k_d K_x \omega_p k_p}{As^2 C_m} \quad \frac{1}{1 + \dfrac{K_x}{As} + \dfrac{K_x \omega_p k_p k_d}{As^2 C_m}(1 + as)}$$

i.e. $\dfrac{\theta_2}{\theta_1}(s) = \dfrac{\omega_0{}^2}{s^2 + \left(\dfrac{K_x}{A} + a\omega_n{}^2\right) s + \omega_0{}^2}$

where $\omega_0 = k_d K_M \dfrac{K_x}{A}$

In this way the damping ratio can be adjusted by altering the constant (a) of the new feedback loop. This additional feedback is known as negative velocity feedback and is frequently introduced for the purpose indicated.

If the effects of leakage and compressibility are now

192

re-introduced, the signal flow diagram is of the form
shown in fig. 6.16 from which the transfer function can
be derived as indicated.

$$\frac{\theta_2}{\theta_1}(s) = \omega_0^2 \Bigg/ \Bigg\{ \frac{JV_0}{BC_m^2} s^4 + \left[\frac{J}{C_m^2} (\lambda_m + \lambda_p) + \frac{JV_0}{BC_m^2} \frac{K_x}{A} \right] s^3 + $$

$$\left[1 + \frac{JK_x}{AC_m^2} (\lambda_m + \lambda_p) \right] s^2 + \left(\frac{K_x}{A} + \omega_n^2 a \right) s + \omega_0^2 \Bigg\}$$

where
$$\omega_0^2 = \frac{K_x k_d k_p \omega_p}{AC_m} = k_d K_L \frac{K_x}{A} = k_d K_L K_M$$

K_M = gain of motor unit

K_L = gain of actuator unit

The reader should establish this transfer function
using fig. 6.16, and then show that if leakage and compressibility are neglected the relationship becomes

$$\frac{\theta_2}{\theta_1}(s) = \frac{\omega_0^2}{s^2 + \left[\frac{K_x}{A} + a\omega_0^2 \right] s + \omega_0^2}$$

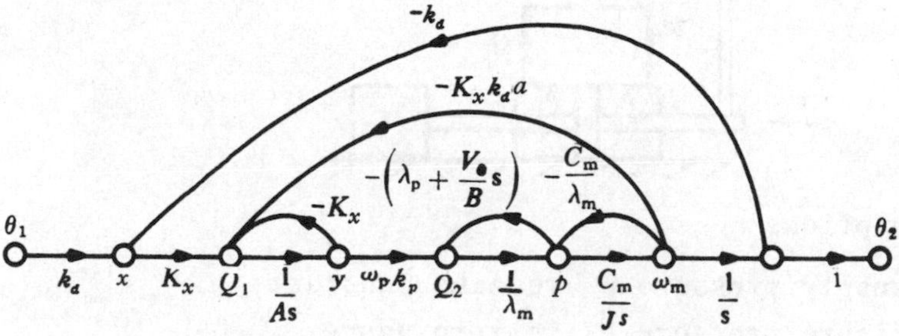

Fig.6.16

VALVE/ACTUATOR SYSTEMS

Revision of linearisation technique: for a valve $Q = C_d a_0 \sqrt{(2\Delta p/\rho)}$, and if a_0 is proportional to spool displacement x then $Q = kx\sqrt{\Delta p}$. For small perturbations

$$\delta Q = \frac{\partial Q}{\partial x} \delta x + \frac{\partial Q}{\partial (\Delta p)} \delta (\Delta p)$$

193

$$= k\sqrt{\Delta p}\ \delta x + \frac{kx\ \delta(\Delta p)}{2\sqrt{\Delta p}}$$

$$= \frac{Q}{x}\ \delta x + \frac{Q}{2\Delta p}\ \delta(\Delta p)$$

$$= K_x\ \delta x + K_p\ \delta(\Delta p)$$

where K_x is the slope of the flow/displacement graph at constant Δp and K_p is the slope of the flow/pressure drop graph at constant x. The ratio K_x/K_p is sometimes referred to as the valve output stiffness.

Position Control using a 5/3 Metering Valve and a Double Acting Symmetrical Actuator

This arrangement is illustrated in fig. 6.17 where feedback is obtained by the link of length $(a + b)$.

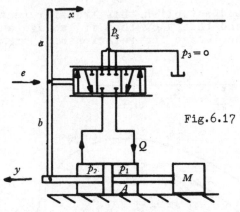

Fig.6.17

Assumptions

(a) supply pressure p_s remains constant
(b) return pressure p_3 is zero gauge
(c) valve flow $Q = ke\sqrt{(\Delta p/2)}$ where Δp = port pressure drop
(d) leakage and compressibility neglected
(e) pure inertia load on actuator
(f) spool dynamics can be ignored
(g) only small displacements are involved

Hence

$$Q = ke\sqrt{\left(\frac{p_s - p_1}{2}\right)} = ke\sqrt{\frac{p_2}{2}} = Asy$$

therefore

194

$$p_s - p_1 = p_2$$

This gives a linearised form as

$$\delta Q = K_e \, \delta e - K_p \, \delta p_1 = K_e \, \delta e + K_p \, \delta p_2 = As \, \delta y$$

where $K_e = |\partial Q/\partial e|_p$ and $K_p = (\partial Q/\partial p)_e$ and, in this case only, are the same for both ports. Therefore

$$\delta Q = K_e \, \delta e - \frac{K_p}{2} (\delta p_1 - \delta p_2) = As \, \delta y$$

But $(\delta p_1 - \delta p_2) A = Ms^2 \, \delta y$

Also $e = \dfrac{b}{a + b} x - \dfrac{a}{a + b} y$ for small movements

therefore

$$\delta e = \frac{b}{a + b} \delta x - \frac{a}{a + b} \delta y$$

From these equations the signal flow diagram in fig. 6.18 can be constructed. From this the transfer function can be established as

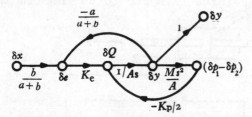

Fig.6.18

$$\frac{\delta y}{\delta x} = \frac{b/a}{\dfrac{a + b}{a} \dfrac{MK_p}{2K_e A} s^2 + \dfrac{a + b}{a} \dfrac{A}{K_e} s + 1}$$

$$= \frac{b/a}{\dfrac{a + b}{a} \dfrac{A}{K_e} s \left(\dfrac{MK_p}{2A^2} s + 1 \right) + 1}$$

The reader should confirm that this transfer function gives the signal flow diagram shown in fig. 6.19, i.e. unity feedback around an integrator (actuator unit), with a single time delay. Note: $K_e/A = K_L$ (steady state gain of the actuator.

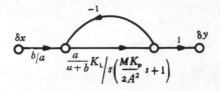

Fig.6.19

The transfer function will also produce the block diagram as shown in fig. 6.20.

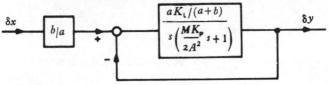

Fig.6.20

WORKED EXAMPLES

3. Fig. 6.21 illustrates a simple position control system where the load actuator is subjected to a force F which always opposes the motion of the piston. Establish the equation of motion for the system when: (a) the force F is constant, (b) the force F is due to an inertia Md^2y/dt^2

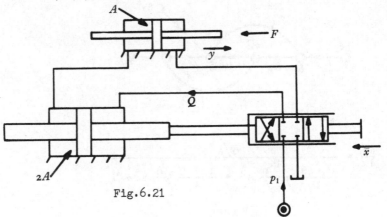

Fig.6.21

and a viscous resistance fdy/dt where y is the movement of the load piston. Assume: oil density = ρ, valve port discharge coefficient = C_d, valve port flow area/unit opening = b.

Flow equations: at the valve the body is moved by the larger diameter actuator. If the load actuator moves y to the right then the other piston must have moved y/2 to the left. Hence valve port opening is $(x - y/2)$. Therefore

196

$$Q = C_d b \left[x - \frac{y}{2} \right] \sqrt{\left(\frac{2\Delta p}{\rho} \right)}$$

where Δp = port pressure drop. But $p_1 = 2\Delta p + F/A$, and at the load actuator $Q = A\,dy/dt$, therefore

$$\frac{A}{C_d b \sqrt{\left(\frac{p_1 - F/A}{\rho} \right)}} \frac{dy}{dt} + \frac{y}{2} = x$$

Note: When motion ceases after a step input at x, the movement of the valve spool and valve body must be the same i.e. $y/2 = x$. Hence the final positional movement of the load actuator will be $y = 2x$ and the equation should be written as

$$\frac{2A}{C_d b \sqrt{\left(\frac{p_1 - F/A}{\rho} \right)}} \frac{dy}{dt} + y = 2x$$

where

$$\frac{2A}{C_d b \sqrt{\left(\frac{p_1 - F/A}{\rho} \right)}} = \tau$$

the time constant of the system, for a constant force F. If the load force F consists of inertia $M\ddot{y}$ together with viscous resistance $f\dot{y}$ then the resulting equation is non-linear and requires a computer solution using arithmetic methods.

However, as in the section dealing with valve controlled transmission systems (speed control), the performance may be studied by considering small movements about a particular operating point. The system equations are

$$Q = C_d b \left[x - \frac{y}{2} \right] \sqrt{\left(\frac{2\Delta p}{\rho} \right)}$$

therefore

$$\delta Q = K_e \left[\delta x - \frac{\delta y}{2} \right] + K_p \, \delta(\Delta p)$$

$$p_1 = 2\Delta p + \frac{F}{A}$$

$$2\delta(\Delta p) = -\frac{\delta F}{A}$$

$$F = (Ms^2 + fs)y$$

$$\delta F = (Ms^2 + fs)\delta y$$

$$Q = Asy$$

$$\delta Q = As\,\delta y$$

A signal flow diagram may be constructed, from the linearised equations, as shown in fig. 6.22. This diagram gives the transfer function as

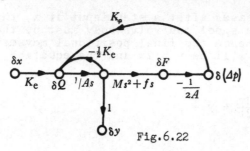

Fig.6.22

$$\frac{\delta y}{\delta x} = \frac{1/2}{\dfrac{1}{\omega_0^2} s^2 + \dfrac{2\zeta}{\omega_0} s + 1}$$

where $\omega_0^2 = \dfrac{K_e}{K_p}\dfrac{A}{M}$

$$\zeta = \frac{1}{2}\left[f + \frac{2A^2}{K_p}\right]\sqrt{\left(\frac{K_p}{K_e AM}\right)}$$

$$K_e = \frac{Q}{x - y}$$

$$K_p = \frac{1}{2}\frac{Q}{\Delta p}$$

It should be noted that the natural frequency (ω_0) developed here is the natural frequency of the complete closed loop control system.

If the transfer function is rewritten in the form

$$\frac{\delta y}{\delta x} = \frac{1/2}{\tau_1\tau s^2 + \tau_1 s + 1}$$

then $\dfrac{\delta y}{\delta x} = \dfrac{1/2}{\tau_1 s(\tau s + 1) + 1}$

198

which gives the block diagram shown in fig. 6.23, i.e. unity feedback around an integrator, with a single time delay.

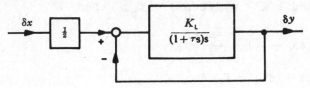

Fig.6.23

4. Establish the transfer function relating small movements of the load mass to small movements of the valve spool for the system in fig. 6.24. Assume p_s remains

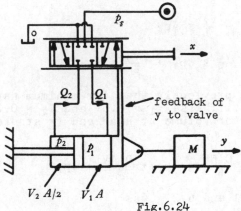

Fig.6.24

constant and neglect connecting line volumes. The fluid bulk modulus is B and the maximum volume of the actuator, at the head end, is V. Assume that, for minimum stiffness, $V_1 = 0.6V$ and $V_2 = 0.2V$.

Flow through valve control ports

$$Q_1 = k(x - y)\sqrt{(p_s - p_1)}$$

and $Q_2 = k(x - y)\sqrt{p_2}$

In the steady state $Q_1 = 2Q_2$, therefore $(p_s - p_1) = 4p_2$.

Linearisation of the flow equation gives

$$\delta Q_1 = k(\delta x - \delta y)\sqrt{(p_s - p_1)} - \frac{k(x - y)}{2\sqrt{(p_s - p_1)}} \delta p_1$$

and $\quad \delta Q_2 = k(\delta x - \delta y)\sqrt{p_2} + \dfrac{k(x - y)}{2\sqrt{p_2}} \delta p_2$

Let $k\sqrt{p_2} = K_e$ and $k(x - y)/(2\sqrt{p_2}) = K_p$; then

$$\delta Q_1 = 2K_e(\delta x - \delta y) - \frac{K_p}{2} \delta p_1$$

and $\quad \delta Q_2 = K_e(\delta x - \delta y) + K_p \, \delta p_2$

therefore

$$\delta Q_1 - 2\delta Q_2 = - \frac{K_p}{2} \delta p_1 - 2K_p \, \delta p_2 \qquad (6.1)$$

Also $\delta Q_1 = As \, \delta y + \dfrac{V_1}{B} s \, \delta p_1$

and $\quad \delta Q_2 = \dfrac{A}{2} s \, \delta y - \dfrac{V_2}{B} s \, \delta p_2$

thus $\delta Q_1 - 2\delta Q_2 = \dfrac{V_1}{B} s \, \delta p_1 + \dfrac{2V_2}{B} s \, \delta p_2$

It has been shown previously that for minimum natural frequency (minimum stiffness) $V_1 = 0.6V$ and $V_2 = 0.2V$ where V = maximum actuator volume at head end ($=$ stroke × A). Therefore

$$\delta Q_1 - 2\delta Q_2 = \frac{0.6V}{B} s \, \delta p_1 + \frac{0.4V}{B} s \, \delta p_2 \qquad (6.2)$$

Combining eqns 6.1 and 6.2 gives

$$- \left[\frac{K_p}{2} + \frac{0.6V}{B} s\right] \delta p_1 = \left[2K_p + \frac{0.4V}{B} s\right] \delta p_2$$

i.e. $\dfrac{\delta p_2}{\delta p_1} = - \dfrac{\dfrac{K_p}{2} + \dfrac{0.6V}{B} s}{2K_p + \dfrac{0.4V}{B} s}$ $\qquad (6.3)$

At the actuator

$$p_1 A - p_2 \frac{A}{2} = Ms^2 y$$

thus $\delta p_1 - \dfrac{\delta p_2}{2} = \dfrac{M}{A} s^2 \, \delta y$

Now $\delta Q_2 = K_e(\delta x - \delta y) + K_p \, \delta p_2 = \dfrac{A}{2} s \, \delta y - \dfrac{0.2V}{B} s \, \delta p_2$

hence

$$K_e(\delta x - \delta y) - \frac{A}{2} s \, \delta y = - \left(K_p + \frac{0.2V}{B} s \right) \delta p_2$$

therefore

$$\frac{A}{2} s \, \delta y + K_e(\delta y - \delta x) = \left(K_p + \frac{0.2V}{B} s \right) \frac{M}{A} s^2 \, \delta y \times$$

$$\frac{1}{\left(\frac{\delta p_1}{\delta p_2} - \frac{1}{2} \right)}$$

therefore

$$\frac{A}{2} s \, \delta y + K_e(\delta y - \delta x) = \left(K_p + \frac{0.2V}{B} s \right) \frac{M}{A} s^2 \, \delta y \times$$

$$\left(- \frac{K_p + \frac{0.3V}{B} s}{3K_p + \frac{0.7V}{B} s} \right)$$

$$\approx - \frac{M}{3A} \left(K_p + \frac{0.3V}{B} s \right) s^2 \, \delta y$$

$$\left[\frac{A}{2K_e} s \left[\frac{MV}{5A^2B} s^2 + \frac{2}{3} \frac{MK_p}{A^2} s + 1 \right] + 1 \right] \delta y = \delta x$$

therefore

$$\frac{\delta y}{\delta x} = \frac{1}{\frac{1}{K_L} s \left[\frac{MV}{5A^2B} s^2 + \frac{2}{3} \frac{MK_p}{A^2} s + 1 \right] + 1}$$

which may be represented by the block diagram shown in fig. 6.25, i.e. unity feedback around an integrator with an oscillatory delay. Loop gain $K_L = 2K_e/A$.

SPEED CONTROL

If it is required to control the output speed of the hydraulic motor in fig. 6.8, then a signal proportional to ω_m is fed back and the input signal will be proportional to speed, say ω_1. If ω is written as the correcting signal then

$$\omega = \omega_1 - \omega_m$$

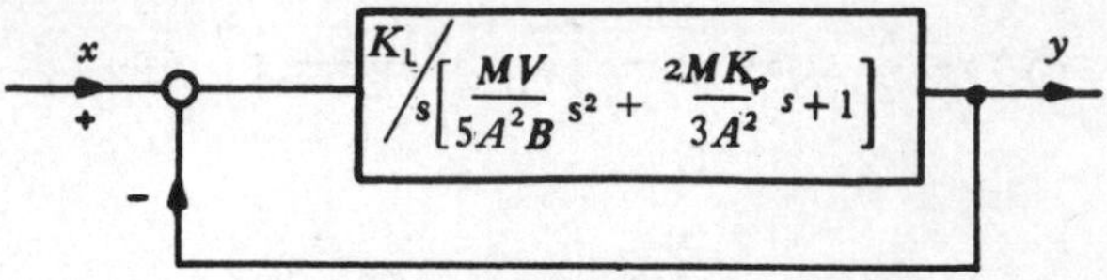

$$\frac{K_L}{s}\left[\frac{MV}{5A^2B}s^2 + \frac{2MK_\varphi}{3A^2}s + 1\right]$$

Fig.6.25

and $C_p = k_p\omega = k_p(\omega_1 - \omega_m)$

Allowing for both leakage and compressibility gives

$$Q = \omega_p C_p - \lambda_p p - \frac{V_0}{B}\frac{dp}{dt}$$

$$= \omega_m C_m + \lambda_m p$$

and $p = \dfrac{J}{C_m}\dfrac{d\omega_m}{dt}$

as before. Hence

$$\omega_m = \frac{\omega_p}{C_m}C_p - \frac{(\lambda_m + \lambda_p)}{C_m}p - \frac{V_0}{BC_m}\frac{dp}{dt}$$

and $\omega_m + \dfrac{J}{C_m{}^2}(\lambda_m + \lambda_p)\dfrac{d\omega_m}{dt} + \dfrac{JV_0}{BC_m{}^2}\dfrac{d^2\omega_m}{dt^2} = \dfrac{\omega_p}{C_m}C_p$

Hence, for zero initial conditions

$$\frac{JV_0}{BC_m{}^2}s^2\,\omega_m + \left[\frac{J}{C_m{}^2}(\lambda_p + \lambda_m)\right]s\,\omega_m + \left(\frac{\omega_p k_p}{C_m} + 1\right)\omega_m$$

$$= \frac{\omega_p k_p}{C_m}\omega_1$$

giving a transfer function of

$$\frac{\omega_m}{\omega_1}(s) = \frac{1}{\dfrac{JV_0}{BC_m\omega_p k_p}s^2 + \left[\dfrac{J(\lambda_p + \lambda_m)}{C_m\omega_p k_p}\right]s + \left(1 + \dfrac{C_m}{\omega_p k_p}\right)}$$

or $\dfrac{\omega_m}{\omega_1}(s) = \dfrac{1}{\dfrac{1}{K_m\omega_n{}^2}s^2 + \dfrac{\tau}{K_m}s + \dfrac{1}{K_m} + 1}$

202

This may be rewritten as

$$\frac{\omega_m}{\omega_1} = \frac{1}{\frac{1}{K_m}\left(\frac{s^2}{\omega_n^2} + \tau s + 1\right) + 1}$$

which gives the block diagram illustrated in fig. 6.26.

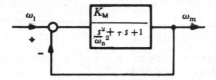

Fig.6.26

If compressibility is ignored the transfer function becomes

$$\frac{\omega_m}{\omega_1} = \frac{1}{\frac{1}{K_m}(\tau s + 1) + 1}$$

which gives the signal flow diagram of fig. 6.27.

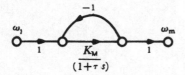

Fig.6.27

If leakage is also ignored then

$$\frac{\omega_m}{\omega_1} = \frac{1}{1 + \frac{1}{K_m}}$$

From the work covered so far it should be noted that systems can be described by transfer functions which are written in terms of specific parameters associated with the systems.

SUMMARY OF PARAMETERS FOR PURE INERTIA LOADED POSITION CONTROL SYSTEM

Motor Gain (steady state)

$$K_M = \frac{\omega_p k_p}{C_m}$$

203

where ω_p = pump shaft speed
C_m = motor capacity/rad
k_p = pump capacity/rad for unit displacement of
stroke mechanism

Motor Natural Frequency

$$\omega_n^2 = \frac{\eta_m}{J} \frac{BC_m^2}{V_0}$$

where η_m = mechanical efficiency
J = polar moment of inertia of load at motor shaft
B = effective bulk modulus
V_0 = trapped fluid volume

Leakage Time Constant

$$\tau = \frac{J(\lambda_p + \lambda_m)}{\eta_m C_m^2}$$

where λ_p = pump leakage rate per unit pressure rise
λ_m = motor leakage rate per unit pressure rise

Damping Ratio

$$\zeta = \frac{\tau}{2} \omega_n$$

Linear Actuator Gain (steady state)

$$K_L = \frac{K_e}{A}$$

Valve Gain

K_e = slope of flow/displacement graph

$$= \frac{Q}{e}$$

where A = piston area
Q = valve flow rate
e = valve port opening (linear port)

Actuator Natural Frequency (double acting - through rod unit)

$$\omega_n^2 = \frac{4BA}{MV^2}$$

204

where M = load mass at actuator
       V = cylinder volume

Leakage Time Constant

$$\tau_2 = \frac{Mc}{A^2}$$

where c = leakage rate per unit pressure differential

Valve Time Constant

$$\tau_3 = \frac{MK_p}{2A^2}$$

where $K_p = \partial Q/\partial \Delta p$

graph $\partial Q/\partial \Delta p$ = slope of flow / port pressure drop

Damping Ratio

$$\zeta = \frac{\tau_2 + \tau_3}{2} \omega_n$$

Differencing Device Gain (usually replaced by transducers and amplifiers in modern systems)

$$k_d = \frac{\text{device output quantity}}{\text{unit difference of inputs}}$$

e.g. $k_d = \dfrac{x}{\theta_p - \theta_m}$

Transducer Gain (steady state)

$$k_t = \frac{\text{output quantity}}{\text{unit input}}$$

e.g. $k_t = \dfrac{V}{\theta_m}$

where V = transducer o/p voltage

Amplified Gain (steady state)

For an electrically signalled system it will be necessary to amplify the correcting signal before applying it to an electro-mechanical device (e.g. torque motor) to obtain an output displacement. Normally such a device will pro- vide a current output as a result of a voltage input.

$$K_A = \frac{\text{output current}}{\text{unit voltage input}}$$

If the loading consists of inertia plus viscous friction then the system equations give, for rotary systems with no feedback loop

$$s\left\{ \frac{JV_0}{\eta_m BC_m^2} s^2 + \left[\frac{J(\lambda_p + \lambda_m)}{\eta_m C_m^2} + \frac{JV_0}{\eta_m BC_m^2}\frac{f}{J} \right] s + \right.$$

$$\left. \left[1 + \frac{J(\lambda_p + \lambda_m)}{\eta_m C_m^2}\frac{f}{J} \right] \right\} \theta_m = K_M \theta_p$$

which may be written as

$$s\left[\frac{1}{\omega_n^2} s^2 + \left(\tau + \frac{f}{J}\frac{1}{\omega_n^2} \right) s + \left(1 + \frac{f}{J}\tau \right) \right] \theta_m = K_m \theta_p$$

i.e. $s\left[\dfrac{1}{\left(1 + \dfrac{f}{J}\tau \right)\omega_n^2} s^2 + \tau\dfrac{\left[1 + \dfrac{f}{J}\dfrac{1}{\tau\omega_n^2} \right]}{\left(1 + \dfrac{f}{J}\tau \right)} s + 1 \right]\theta_m$

$$= \frac{K_m}{1 + \dfrac{f}{J}\tau}\theta_p$$

i.e. K_m is now $\dfrac{K_m}{\left(1 + \dfrac{f}{J}\tau \right)}$

ω_n^2 is now $\omega_n^2\left(1 + \dfrac{f}{J}\tau \right)$

τ is now $\tau\left(1 + \dfrac{f}{J}\dfrac{1}{\tau\omega_n^2} \right)$

ζ is now $\dfrac{\zeta\left(1 + \dfrac{f}{J}\dfrac{1}{\tau\omega_n^2} \right)}{\left(1 + \dfrac{f}{J}\tau \right)^{\frac{1}{2}}}$

for linear systems with no feedback loop

206

$$s \left\{ \frac{1}{\omega_n^2} s^2 + \left[\tau_2 + \tau_3 + \frac{f}{M\omega_n^2} \right] s + \right.$$

$$\left. \left[1 + \frac{f}{M} (\tau_2 + \tau_3) \right] \right\} \delta y = K_L \, \delta x$$

from which similar variations of system parameters can be established.

FURTHER EXAMPLES

1. The simple hydraulic servosystem shown in fig. 6.28 is used to raise and lower equipment, of total mass m, over a small distance from the surface plate. Establish the equation governing the system and calculate the value

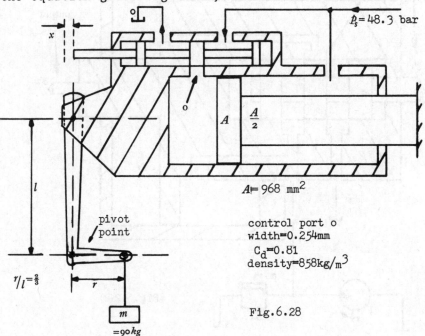

$p_s = 48.3$ bar

$A = 968$ mm^2

control port o
width=0.254mm
$C_d = 0.81$
density=858kg/m^3

Fig.6.28

of the time constant. Explain the significance of this term. Determine also the time taken for the system error to reduce to 1% of its initial value after being subjected to a step position input at the valve spool. Neglect leakage, compressibility and friction.

$$(\tau s + 1) Y = X \text{ where } \tau = A \ C_d' a_0 \ \frac{2}{\rho} \left(\frac{P_s}{2} - \frac{mgr}{Al} \right)^{-1} ;$$

$$t = 0.312 \text{ s}$$

2. Fig. 6.29 illustrates a position control servomechanism where friction and inertia effects are to be neglected and the load is zero. T_1 and T_2 are time constants associated with motion to the right and the left respectively. Prove that

$$\frac{T_1}{T_2} = \left[\frac{A_2}{A_1} - 1\right]^{\frac{1}{2}}$$

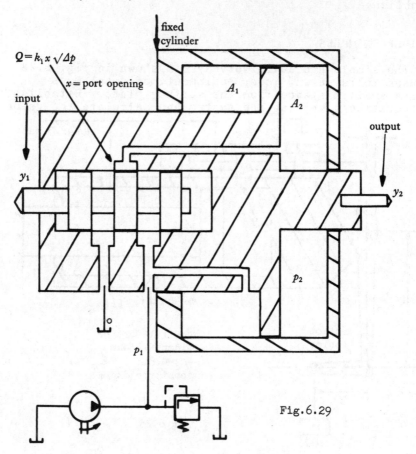

Fig.6.29

Assuming now that $A_2 = 2A_1$, show that, for a step input displacement of 2 mm, the output after T_1 seconds will be 1.264 mm. What will be the output after 10 T_2 seconds?

3. A simple position control servosystem is illustrated in fig. 6.30. The three-way valve has a rectangular control port with underlap of an amount u. The supply gauge pressure to the valve is constant at p_s and the tank port pressure may be taken as zero gauge. The ram piston area is twice the rod area and load, leakage and compressibility

208

effects are to be ignored. If the system time constant is T_1 when $u = 0$ and is T_2 when $u > 0$ show that $T_1 = 2T_2$ assuming spool movement $x < u$.

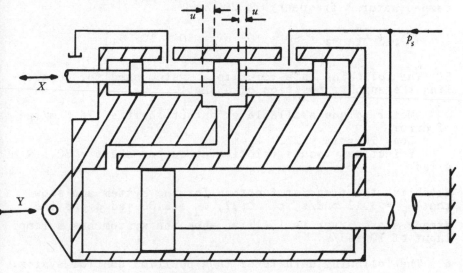

Fig.6.30

Explain the meaning of the term 'time constant' assuming the system is subject to a step displacement input x at the valve spool where $x < u$.

4. A simple hydraulic position control system is illustrated in fig. 6.31. θ_1 and θ_2 are input and output angular displacement (radians). The pure inertia load J is

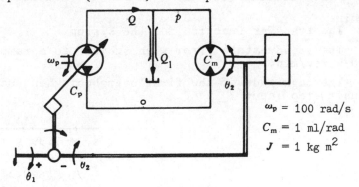

$\omega_p = 100$ rad/s
$C_m = 1$ ml/rad
$J = 1$ kg m^2

Fig.6.31

driven by a fixed-capacity hydraulic motor. The variable-capacity pump has a capacity

$$C_p = 2.5 \ (\theta_1 - \theta_2) \ \frac{ml}{rad}$$

The oil leakage rate from the high pressure side of the system is equivalent to Q (leakage) = $0.5 \times 10^{-3}p$ ml/s (where p is the pressure in bars). Establish the differential equation governing the system and show that the damped natural frequency is 200 rad/s.

$$\left(\ddot{\theta}_2 + \frac{10^3}{5} \dot{\theta}_2 + 50 \times 10^3 \, \theta_2 = 50 \times 10^3 \, \theta_1 \right)$$

5. The following data refer to a system used for controlling the angular position of a shaft.

Motor torque available at output shaft = 218 N m/rad of error
Load inertia = 22.4 kg m^2
Effective viscous friction at output shaft = 30.5 N m s/rad

Establish the transfer function for the system and show that ω_n = 3.12 rad/s, ζ = 0.22, ω_d = 3.05 rad/s and the steady-state error is 1.41 rad when the system has a ramp input of 10 rad/s.

6. The following data refer to a position control system.

Error-detecting device gives 1 V/rad of error between input and output shafts
Amplifier gain = 100 mA/V input
Servo motor output torque = 10 Nm/A of field current
Moment of inertia of motor armature = 25×10^{-3} kg m^2
Moment of inertia of load = 50×10^{-2} kg m^2
Speed reduction gearbox between motor and load is 5:1
Viscous friction at load shaft = 1.0 N m/(rad/s)

Determine
(a) The transfer function for the system.

(b) The steady-state error when subject to a ramp input of 300/π rev/min.

(c) The magnitude of the first overshoot when subject to a unit step input.

$$\left(\frac{\theta_o}{\theta_i}(s) = \frac{4.44}{s + 0.89 \, s + 4.44} \right);$$

0.32 rad ;

0.26 rad).

7. Fig. 6.32 shows a hydraulic relay designed to provide an output displacement y proportional to input force F. The mass of the valve spool is M and its motion is opposed by a viscous damping force equal to c times the velocity,

and a spring of stiffness k attached between output and input as shown. The output velocity is K_L times the valve spool displacement. Show that the transfer function for the relay is

$$\frac{Y}{F}(s) = \frac{1}{\dfrac{1}{K_L} \, s \, (Ms^2 + cs + k) + k}$$

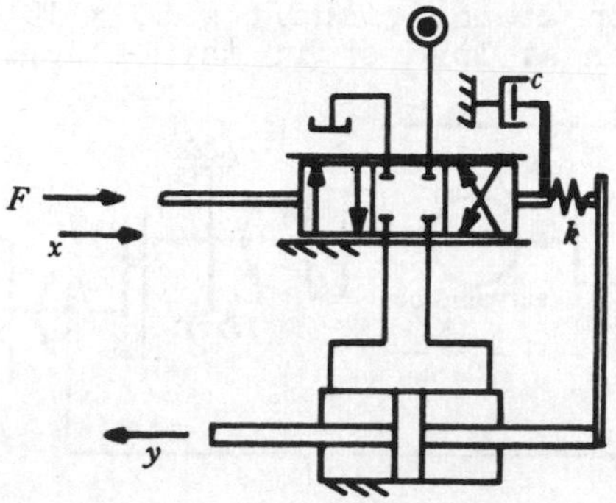

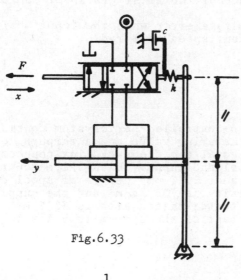

Fig.6.32

8. The hydraulic relay of question 7 is slightly modified as shown in fig. 6.33. Show that the transfer function is

Fig.6.33

$$\frac{Y}{F}(s) = \frac{1}{\dfrac{1}{K_L} \, s \, (Ms^2 + cs + k) + 2k}$$

and that for stability $c > 2MK_L$.

9. A servosystem is indicated in fig. 6.34 and the following data apply

Servomotor moment of inertia = 2×10^{-6} kg m^2 (J_m)

Load moment of inertia = 5×10^{-3} kg m^2 (J_L)

Motor shaft speed/load shaft speed = 100/1

Servomotor torque output per radian of controller
input = 50×10^{-3} N m (K)

Viscous friction at motor shaft = 50×10^{-6} Nm/rad s^{-1}

Viscous friction at load shaft may be neglected.

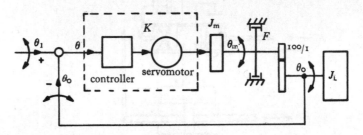

Fig.6.34

Establish the differential equation for the system and determine
(a)   the natural damped frequency.
(b)   The magnitude of the first overshoot in response to
a unit step input.
(c)   the steady-state error when the input shaft is ro-
tated with a constant speed of 2 rad/s.

(10 rad/s;

0.044 rad;

0.2 rad)

10. A simple valve-controlled servosystem consists of a
5/3 closed centre metering valve and a through rod actuator. There is an equal arm feedback link connecting actuator ram, valve spool and input. Each valve control port
orifice has an area of 1.27 mm^2 per mm of spool displacement. The oil density is 706 kg/m^3 and the supply pressure
is 69 bar. The actuator piston area is 3225 mm^2. Show
that the time constant of the system is 5.1×10^{-2} s.